FIGHTER NIGHTS

Banner Books is Australia's specialist publisher of military history, committed to publishing high-quality books which add a significant dimension to the history of Australians at war.

Also available:

The Decisive Factor: 75 & 76 Squadrons – Port Moresby and Milne Bay 1942 by David Wilson

Lion over Korea: 77 Fighter Squadron RAAF 1950-53 by David Wilson

Six Aces: Australian Fighter Pilots 1939-45 by Lex McAuley

Tocumwal to Tarakan, Australians and the Consolidated B-24 Liberator by Michael V Nelmes

Strike and Strike Again 455 Squadron RAAF 1944-5 by Ian Gordon

FIGHTER NIGHTS

456 Squadron RAAF

JOHN BENNETT

BANNER BOOKS

First published in 1995 by Banner Books
PO Box 937
Belconnen ACT 2616
Australia

Production by Sylvana Scannapiego
Island Graphics Pty Ltd
Typesetting and Design by Island Graphics Pty Ltd
Jacket design David Constable
Printed in Australia by Impact Printing Brunswick

Cataloguing-in-Publication information:

Bennett, John (John William), 1949- .
Fighter Nights : 456 Squadron RAAF

Bibliography.
Includes index.
ISBN 1 875593 10 1.

1. Australia. Royal Australian Air Force. Squadron, 456 - History. 2. World War, 1939-1945 - Aerial operations, Australian, 3. World War, 1939-1945 - Regimental histories - Australia. I. Title.

940.544994

CONTENTS

PROLOGUE

A "fighter night" was originally the terminology for a bright night, when the defending anti-aircraft guns would be silenced, and the air defence of England was entrusted to patrolling flights of Defiants and Hurricanes, relying on good eyes, searchlights and moonlight. The tactics of these "cat's-eyes" fighters had differed little from the First War.

As technology developed, airborne intercept radar, assisted by ground controllers, overcame the handicap of darkness. Night fighter crews could stalk, identify and destroy their prey. These advances led to much night fighter success, against enemy aircraft over England and over Germany leading to every night becoming a "fighter night".

This is the story of the only Australian night fighter squadron - No 456 Squadron. The members of 456 had come to help Britain like their fathers had done in 1914-18. They were fighting a war on the leading edge of radar technology, the likes of which were not to be experienced by their colleagues in the Pacific war.

FOREWORD

One of the great privileges of my wartime experiences was to have been a member of, and subsequently acting CO, of 456 Night Fighter Squadron RAAF.

Not only was the Squadron the only night fighter unit of the RAAF in World War Two, but it distinguished itself by being the second highest scorer of enemy aircraft of all RAAF squadrons. From its early beginnings in Wales in 1941 flying Boulton Paul Defiants, then Beaufighters and Mosquitos, the Squadron played a significant role in the defence of Great Britain. Its success prior to, during and after the Normandy landings was a credit to the aircrew and groundcrew alike. In no other squadron in which I served was there such a closeness between all members of the Squadron which lives to this day, and is evident at all our reunions where there continues deep friendships which have little or no bearing to rank, trade or capacity in which they served.

Memories of those days are filled with a wide range of emotions. The dry-mouthed feeling on Intruder missions far over enemy territory; the responsibility of correctly recognising an aircraft in the blackness of night; the awful weather encountered; the boredom, at times, on patrol; the wonderful parties to relieve tension; the sadness at losing a friend; the cramp and cold of the cockpit; the utter reliance on our instruments; the joy of deep enduring friendships; and last, but not least, the pride in being part of a team that played a significant role in our winning the war in the air, at night in particular.

We are all indebted to John Bennett for his painstaking research in compiling this history which, together with our works of a similar nature, will form part of our Nation's history to be passed on to future generations.

I would like to pay a special tribute to just a few of the many members of our Squadron who played significant roles leading to our success.

Wing Commander Gordon Olive DFC assumed the responsibility of being the Squadron's first CO, and was a Battle of Britain veteran, one of "The Few", and a distinguished Australian. The late Wing Commander Dr Frank

Parle, who as a father figure and long serving member of the Squadron, was looked up to with great respect. Squadron Leader Stewart Green OBE, in his capacity of Engineering Officer with his team of fine NCOs and men; set an example and standard of serviceability on which our lives depended, and was the envy of other squadrons. Wing Commander Keith Hampshire DSO & Bar, DFC, whose drive and enthusiasm contributed to its success in 1944. The quiet, dependable and greatly respected Wing Commander Basil Howard DFC, who tragically died at the war's end and was a terrible loss to us all.

I would like to pay tribute to my fellow Squadron members and remember especially those who died during the war and since, and I am sure that they, like me, would be grateful to John that the history of our Squadron has been so faithfully recorded.

R B Cowper

ACKNOWLEDGMENTS

Any book that attempts to trace the story and life of a squadron must rely, in the first instance, on the documentation of the facts and figures of the unit. For that, the researcher starts at the RAAF Historical Section in Canberra, which keeps the Operations Record Books for RAAF units. I therefore thank David Wilson and his staff for the services they provide. For the personnel biographical details, many hours of research are necessary at the RAAF Personnel Records section at Queanbeyan and I am indebted to David Pullen and his staff for keeping these records available. Also, I am grateful to the RAF Air Historical Branch, London, for the records they keep on RAF aircraft.

That sets the framework for a history, but what is necessary to complete the story is the cooperation of those who were there. I thank the members of the 456 Squadron Association for the wholehearted support that they have offered during this project.

From the 456 Squadron aircrew, much time was spent with Fred Stevens, Col Griffin and Gordon Gatenby, all of whom also had a good photographic coverage from their service on 456. In addition, the assistance of Rex Timbury and David Spring, and the letters and photographs from some of the groundcrew, Len Stamp and Ron Vidler, is greatly appreciated and provided many necessary details. I thank Graham Walton and Tony Bennett for their photographic assistance. My gratitude also to Mrs Judith Wolfe, and to Group Captain Dave Schubert for his epilogue.

David Vincent, a fine aviation historian in his own "write", was generous with time, advice and photographs, and this assistance was crucial in allowing me to complete this work in time (that most precious of all resources!).

Bob Cowper DFC & Bar, the last Commanding Officer of No 456 Squadron, was kind to accept my offer of writing the Foreword to this book. I hope that I have done him, and the other members of 456, justice in this, their record.

Lex McAulay of Banner Books was willing to take the chance in publishing this book, and I hope it is a fitting companion to the work he has already published on No 455 Squadron.

Finally, a project of this magnitude is an all time-consuming task. I hope my family, Marijke, John and Rob, can understand my lapses of attention that have been dictated by a compelling need to see that stories such as this are told.

CHAPTER ONE

FIGHTER COMMAND

The euphoria of the post-1918 era in Britain had brought the assumption that the British Empire would not likely be involved in any major war for the ensuing ten years. This "Ten Years' Rule" became financially convenient for successive governments to avoid the cost of any defence preparedness. Indeed, it was felt the only future aggressor would be France – the only major European country still capable of maintaining strong military and naval forces. The defence of the United Kingdom was considered completely safe in the hands of the army, and in Britain's traditional first-line of defence, the Royal Navy.

By 1934, it had been realised that the ultimate potential enemy would be Nazi Germany. The General Staff believed that Germany might be ready for a major conflict by 1938, or at the latest 1939. Intelligence had already gathered ample evidence of Hitler's secret air force, the *Luftwaffe* – built in open defiance of the restrictive Versailles Treaty – which, by the end of 1934, had at least 22 first-line squadrons of bombers and fighters[1]

The resulting preparations for expansion of the RAF in 1935 led to several significant issues. These involved a reorganisation for the expanded service, the acquisition of new aircraft with the need for increased aircraft production, and parallel to this was the development of radar. In addition, a factor which was to have immense impact on Australia was the provision of personnel.

Since 1925, the major RAF Command had been the Air Defence of Great Britain (ADGB), which encompassed all bombers, fighters and air defences (anti-aircraft guns and searchlights). The foundation for winning the Battle of Britain was laid in 1934 when the threat of Germany was taken seriously

1 C Bowyer, *Fighter Command*, Sphere, London, 1981, p.7.

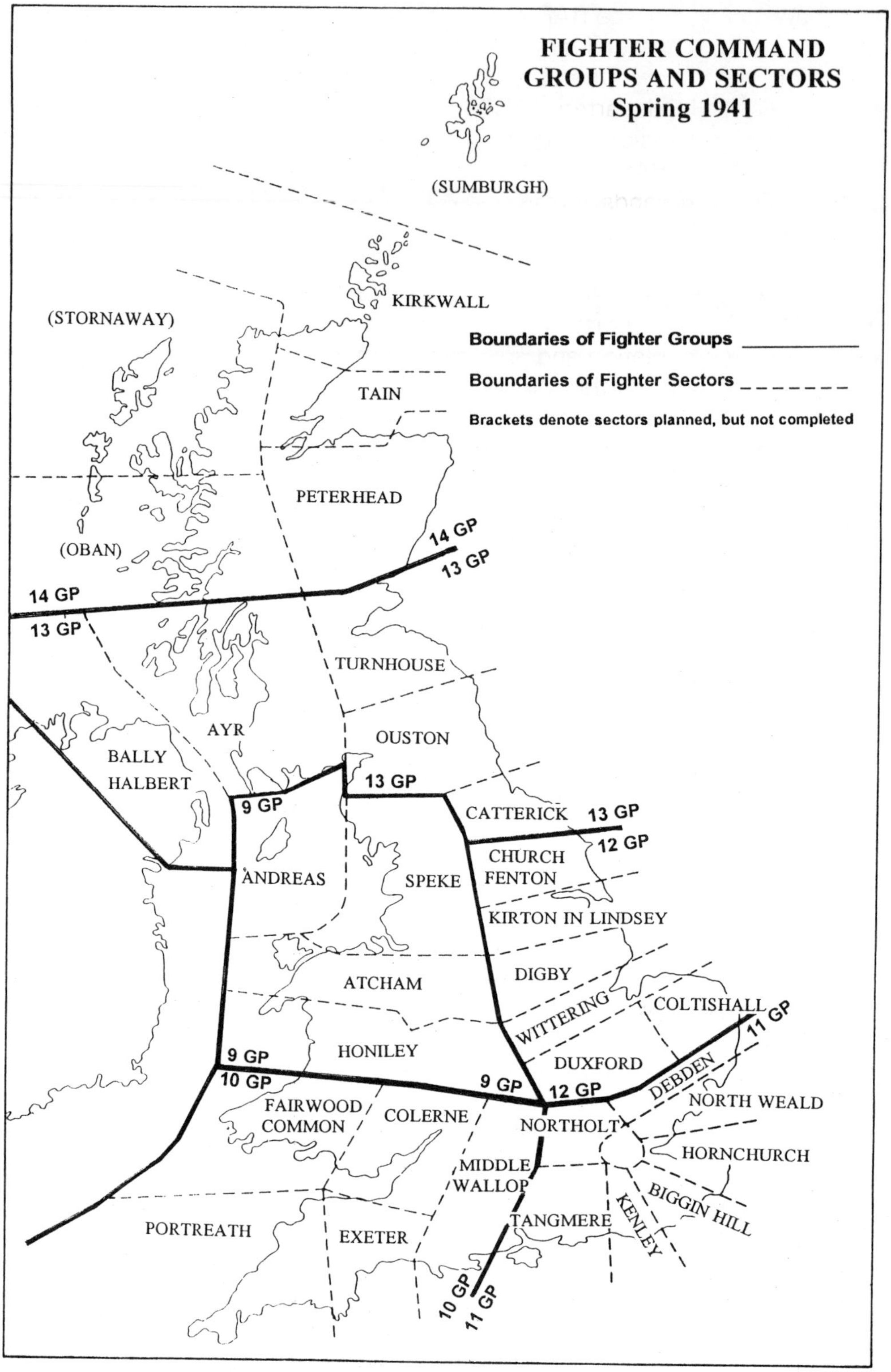
FIGHTER COMMAND
GROUPS AND SECTORS
Spring 1941
Boundaries of Fighter Groups
Boundaries of Fighter Sectors
Brackets denote sectors planned, but not completed
(SUMBURGH)
KIRKWALL
(STORNAWAY)
TAIN
PETERHEAD
(OBAN)
14 GP
13 GP
TURNHOUSE
AYR
BALLY HALBERT
OUSTON
9 GP
CATTERICK
12 GP
ANDREAS
SPEKE
CHURCH FENTON
KIRTON IN LINDSEY
ATCHAM
DIGBY
WITTERING
COLTISHALL
11 GP
HONILEY
DUXFORD
10 GP
DEBDEN
FAIRWOOD COMMON
COLERNE
NORTHOLT
NORTH WEALD
HORNCHURCH
MIDDLE WALLOP
BIGGIN HILL
TANGMERE
KENLEY
PORTREATH
EXETER

and the expansion of the RAF was approved. The following year, plans were put into effect to build an efficient air defence system around radar detection, the observer corps network, and centres for the control of fighters under one Headquarters at Bentley Priory, near London.

The 1935 expansion program presented difficulties under the ADGB structure, and the RAF was drastically reorganised with the creation in 1936 of four new commands formed on a functional basis rather than regional: Fighter, Bomber, Coastal and Training Commands. At the time of Dunkirk in 1940, there were three fighter groups – Nos 11, 12 and 13. By the opening of the Battle of Britain, No 10 Group was operational in the south-west, and two others (No 9 in north-west England and No 14 in northern Scotland) were nearing completion and approaching an operational state.

The Groups were further divided into Sectors, each containing a number of fighter bases, or stations. One station was nominated as a Sector Station, from where the Sector Controller, in his operations room, controlled his fighters. The Group operations room had a broader picture of its Sectors, and allocated squadrons to counter particular raids, while Fighter Command's operations room organised reinforcements amongst the Groups.

Until the second half of the 1930s, RAF fighters were open-cockpit biplanes, little advanced in design from their predecessors of 1918, armed with twin guns of rifle calibre (Bulldog, Demon, Gauntlet). Immediate future replacements envisaged were also biplanes (Gladiator, Fury II). Air Ministry Specification F5/34, issued in January 1935, had called for a high speed monoplane fighter, capable of destroying a bomber with just two seconds' gunfire. This required four or six machine guns, but was later raised to eight wing-mounted .303 guns. The Air Officer Commander-in-Chief ADGB, Air Chief Marshal Sir Robert Brooke-Popham, perhaps best known for his later role in the Singapore debacle, expressed the view "...eight guns was going a bit too far, I should have been content with four".[2] Brooke-Popham further showed his appreciation of fighter development by stating the guns should be fitted to the cockpit area, and expressed his opposition to enclosed cockpits. It is not recorded whether he was against retractable undercarriage.

Fortunately these views were overridden, which led to the Hurricane and the Spitfire, high speed designs which had benefited from the Schneider Trophy air races of the early thirties. Both were armed with eight wing-mounted Browning machine guns, each supplied with approximately 300 rounds of ammunition - sufficient for 15 seconds firing. These fighters were able to enter RAF service over the second half of 1938.

However, by September 1938, to oppose the *Luftwaffe*, Fighter Command could only muster 93 of the new eight-gun fighters. All the remainder of its 666 aircraft were outdated biplanes. Aircraft output rose steadily from 158 aircraft per month in April 1938, until nearly 800 per month at the outbreak of war, which equalled that of the Germans.[3] 1938

2 loc cit, p.18.

3 D Richards, *Royal Air Force 1939-1945, Vol I*, HMSO, London, 1953, p.17.

and 1939 were to provide a breathing space, as even these front line Hurricanes and Spitfires had no constant-speed propellers, no bullet-proof windscreens, no armour plate, no gun heating, and no explosive ammunition. Furthermore, the radar chain was incomplete, with only five radar stations working out of the twenty required. The RAF was far from ready to go to war; the respite offered by the Munich agreement allowed the time to build up and modernise the RAF.

Lord Beaverbrook was appointed by Churchill in May 1940 as Minister of Aircraft Production, and immediately instituted an innovative aircraft production system, with fighters as the priority. This enabled the RAF squadrons to maintain their strength over the forthcoming difficult months.[4] This rate of production of fighters was now twice that of the Germans.

For the fighters to intercept any attacking force, information was gained from two sources – the new radar stations covering an area outside the British Isles, and the Observer Corps, which would trace the passage of aircraft over land. In 1935, the Radio Department of the National Physical Laboratory supplied details of the general principles of radio direction finding (RDF), and how an RDF chain could be established for operational use for the RAF. The Air Member for Research and Development was Air Marshal Sir Hugh Dowding, who not only supported this work, but in 1936 enthusiastically took the plans to his new post as AOC-in-C Fighter Command. He described this development as "a discovery of the highest importance" and as a "tremendous leap forward".[5] RDF – which became known as radar (radio direction and ranging) – was able to provide information on the incoming raids to allow interception by defending fighters. Under Dowding, a chain of twenty RDF stations, with their 80-metre high masts, was commissioned, which provided the nucleus of a much longer chain of units stretching along England's southern and eastern coastlines, all operational by early 1940.

Dowding's foresight and policies were to have a huge bearing upon the state of Britain's air defences by the outbreak of war in 1939. Radar enabled the RAF to concentrate its force into a comparatively small area. However, the radars were only effective over the sea, and as raids crossed the coast they were tracked visually by the Observer Corps. Churchill was to describe this handover from radar to the Observer Corps as "a transition from the middle of the twentieth century to the early Stone Age".[6]

Personnel shortages were initially to be addressed by an effective use of

4 British fighter production over 1940 was impressive (Richards, op cit, p.152):

Month	Planned Fighter Production	Actual Fighter Production
Feb 1940	171	141
May 1940	261	325
Jun 1940	292	446
Jul 1940	329	496
Aug 1940	282	476

5 Bowyer, op cit, p.88.

6 AVM J E Johnson, *The Story of Air Fighting*, Hutchinson, London, 1985, p.155.

The "Chain Home" radar towers along England's south-east coast were the eyes of the whole defensive system during the Battle of Britain. They could detect the approach of enemy raiders when they were still 60 miles or more away. RDF could also determine direction and speed, but could only approximate force strength and height.

Reserve forces. With the ominous threat to peace, the RAF Volunteer Reserve had commenced in April 1937, and by the outbreak of war, 5,000 young men of the Volunteer Reserve had completed, or were undergoing, training as pilots. However, this was far from enough. The vast expansion of the RAF was to rely on manpower from the Commonwealth, or the Dominions.

On 17 December 1939, the Empire Air Training Scheme (EATS) agreement was signed in Ottawa to enable the training of Dominion aircrew as a common enterprise. Under the terms of the agreement, the UK would supply nearly all the aircraft and a nucleus of skilled men, the Dominions all other requirements. Australia agreed to supply 36% of the Dominions' EATS aircrew.[7] By mid-1942, when the EATS organisation would reach full size, it was to be capable of producing no less than 11,000 pilots and 17,000 other aircrew each year.[8]

As part of Article XV of this agreement, the Dominions undertook to form national squadrons for operations within the RAF. Such units would be manned as much as possible by that country's nationals, with a national identity, and if possible, commanded by a member of the respective air force. However, operational and administrative control would remain with the RAF. Eighteen squadrons were planned to be formed by the RAAF –

7 J McCarthy, *A Last Call of Empire*, AWM, Canberra, 1988, p.21.

8 Richards, op cit, p.73.

Nos 450 to 467 Squadrons. The terms and conditions of these Article XV squadrons, as they became known, were to be confusing due to differing perspectives of the EATS agreement. This was to remain a bone of contention by many who served on Article XV squadrons throughout the war.

Another training problem addressed was that of the operational training of aircrew. Operational Training Units (OTUs), large organisations of up to 70 aircraft on a base, were formed from 1940 to provide the final stages of aircrew training on the operational types they would fly on squadrons. This system of training was to be adopted throughout all the operational Commands, and proved the key to efficiency in the air.

From the beginning of the *Blitzkrieg* through the Low Countries and France in May 1940, until the British withdrawal at Dunkirk nearly a month later, the RAF lost no less than 432 Hurricanes and Spitfires.[9] Dowding's fear of being bled of fighters was coming true. Moreover, the many pilots killed in action were the most grievous losses. These men had become veterans in mere weeks of action, and their experience was now lost for ever to Fighter Command's squadrons. However, the *Luftwaffe's* mission at Dunkirk to prevent the evacuation of British and French troops had failed. Bad weather, lack of forward airfields and poor training in pin-point bombing had denied its success. During the nine days of the British evacuation, the *Luftwaffe* could only operate in strength for 2 1/2 days, and that was in the face of stout RAF fighter cover, now able to operate from bases which were relatively close.

The Battle of Britain is regarded to have begun on 10 July 1940. The real attack - *Adler-Tag* or Eagle Day – had been fixed by Goering for 13 August. The *Luftwaffe's* objective was to subdue Britain by air power alone. The primary target was the RAF, driving it out of the skies as rapidly as possible, and destroying its ground organisation of airfields and coastal radar stations, and the industry that fed it. The goal was therefore clear – the subjugation of the RAF was the first and foremost aim, and with that the elimination of Fighter Command. The Germans planned that a few hard blows over three consecutive days would so reduce the RAF that control of the air – the precursor for subsequent operations – would be won.

Weather prevented the main fighter bases being attacked in these first days, and mistakes – poor intelligence and navigation errors on Eagle Day – led to wrong airfields being attacked. In addition, continuous attacks were not coordinated, which allowed the defending Spitfires and Hurricanes to refuel and re-arm. Any German crew shot down was over hostile territory, unlike Fighter Command pilots who baled out and were hopefully available to fight again. Moreover, the Germans had grossly underestimated Britain's air strength.

From the *Luftwaffe's* perspective, the battle had been lost in 1938 because the German participants believed it would never happen. Production of a long-range, heavy, four-engined bomber was not

9 loc cit, p.145.

The Ju 88 proved to be a disappointment as a bomber, and this and the other existing bombers proved slow, vulnerable and too light.

contemplated, priority instead going to concentrate resources to the Junkers Ju 88 dive bomber. This strategy of no heavy bomber had proved correct for the *Blitzkrieg* through the low countries, but now in the summer of 1940, the *Luftwaffe* faced quite a different campaign. Suddenly the gap in its equipment was revealed.

On 7 September, the *Luftwaffe* was ordered to switch its target to London, to use the capital as bait and lure Fighter Command to battle. This was the fundamental German mistake. With no attacks on its airfields, the RAF was given time to replenish its fighter defences. Dowding was able to repair his damaged airfields and reorganise and resupply his mauled squadrons. In just a week, No 11 Group was returned to full strength. Hitler's vengeance attacks on the British capital probably saved Fighter Command from near extinction.

The following week, on 15 September – which the RAF remembers as Battle of Britain Day – the *Luftwaffe* found that the fighter defences appeared stronger than ever, as the RAF had committed all its reserves. The Germans were being hampered by the close escort tactics of their fighters, which tied them to the slow bombers, and, unable to exploit their flying attributes, were in a poor position to repel RAF fighters.

On average, the RAF was to lose two fighters for every German one.[10] Some historians therefore query whether the Battle of Britain was indeed an RAF victory. Perhaps it is on the face value of these statistics they argue their case. But it is undeniable at the strategic level, that the victory was indeed Fighter Command's, as "Operation Sealion" – the proposed German invasion of Britain – never eventuated, because German control of the air over south-east England had never been achieved. The *Luftwaffe* had tried to prove for the first time the dictum of the Italian theorist General Giulio Douhet, that future wars could be won by strategic air attacks alone. The *Luftwaffe* had lacked the power to do so. The RAF had met its "finest hour".

Between 1 August and 1 October 1940, the two *Luftwaffe* Air Fleets engaged in operations against Britain lost 500 aircraft, and the *Luftwaffe* had been defeated in its attempts to bomb Britain out of the war by day, and turned to night raids. London was subsequently attacked by *Luftwaffe* bomber aircraft numbering between 100 and 300 at a time. The darkness protected them and bestowed control of the air that they never achieved by day. For Fighter Command, the Battle of Britain was over, the Battle of the Night was about to begin.

10 Official RAF figures for Aug 1940 give 390 Spitfires/Hurricanes destroyed, 197 badly damaged. Luftwaffe official figures give 231 German fighters lost, 80 damaged, including losses on the ground. C Bekker, *The Luftwaffe War Diaries*, MacDonald, London, 1966, p.170. However, it must be remembered that RAF fighters were going for the bombers, not the fighters.

CHAPTER TWO

NIGHT FIGHTING

The concept of aircraft fighting at night was not new. During the First World War, as early as 1916, dedicated scout units of the Royal Flying Corps on home defence duties had been flying at night. Over 1917 and 1918, these night fighter squadrons had engaged the Zeppelin airship raids and the giant Gotha bombers, which roamed against London, the Midlands and northern England.

No 151 Squadron, of the newly-formed Royal Air Force, had been established in June 1918, and was soon operational providing night defence over France, in the Abbeville area. Sopwith Camels pioneered tactics that were to be used to great effect in the Second World War, and took the offensive, bombing German airfields by night and machine gunning any aircraft attempting to fly. The Camels also flew bomber support missions by attacking enemy searchlights which threatened to light up the British bomber force.

Now, in this new war, the Germans had again switched to night offensive operations. Between 7 September and 13 November 1940, more than 12,000 night sorties were flown over Great Britain in a concentrated offensive against London. The British night defences could only claim 81 of the raiders: 54 by guns, 8 by night fighters, 4 by balloons, and the remainder by other causes.[1]

On the night of 14 November 1940 - two months after the *Luftwaffe* realised the Battle of Britain had been lost - the Germans bombed Coventry. Utilising a radio beam to guide twelve Heinkel He 111s over the target, they had also used transmitted crossing beams that had enabled them to calculate wind, ground speed and distance. Marking the target at Coventry

1 Richards, op cit, p.206.

by flares, the main attacking force of 400 German bombers then unleashed high explosive and incendiary bombs and mines on the city for the next few hours. The attack on this moonlit night caused devastation and industrial damage that shocked the British public - mainly because it was realised that little could be done against such night terror attacks.

The British soon deduced that the *Luftwaffe* was using a radio beam system to enable blind bombing over England. Development had in fact started in 1934 to develop radio directional navigation, which enabled a bomber crew to fly "blind" along the a directional beam and drop bombs with relative accuracy. Transmitted from stations in western Germany and the occupied European coast, the *Knickebein* ("crooked leg") system employed two beams in the very high frequency (VHF) band, on frequencies between 30 and 33 MHz. A pilot would fly along one of the beams aurally, by means of a series of dots and dashes, using standard Lorenz blind flying equipment fitted to the entire German bomber force. Upon intersection with a second beam, 20 kilometres (twelve miles) from the target and indicated by a further tone, a timed run was accurately flown to compensate for the wind.[2] *Knickebein's* short wave beam was no more than a kilometre wide over London, which allowed a bombing accuracy to within roughly a square mile to be achieved.[3] This was sufficient for the mass urban bombardment of the Blitz.

To counter the *Knickebein* effectiveness, the British began interfering with the German beams. Under the code name "Headache", the enemy signals were received, and then retransmitted, to broaden the beam and thereby cause a loss of accuracy by deviation from the desired ground track. In addition, high powered jammers, code-named "Aspirins", were quickly developed over the summer of 1940 to transmit signals to confuse the aural steering directions, enabling *Knickebein* to be effectively countered by September 1940.

The next German beam system was fitted only to its Pathfinder Force - *Kampfgruppe 100*, KGr100. This system, known as *X-Gerät*, or X-Apparatus, enabled the crew to accurately fly an approach beam aurally and visually, by following a direction indicator, which like *Knickebein* had been laid over the target. This beam was no wider than 20 metres over typical targets in the Midlands, which was 200 miles from the main beam transmitter. As the target was approached, a forewarning signal beam crossed the main beam to indicate 20 kilometres to run. A second cross beam, the main cross signal, was five kilometres before the target. The time to cover the 15 kilometres between the two cross beams was automatically calculated by a small mechanical computer, known as the X-Clock, to calculate ground speed. With this information, the mechanism was then able to compute an

2 The aircraft ground speed was calculated over the first six miles, which enabled an automatic countdown for bomb release over the remaining distance. Bekker, op cit, p.180.

3 R V Jones, *Most Secret War*, Hamish Hamilton/Coronet, London, 1979, pp.126, 179; Richards, op cit, p.199.

accurate automatic bomb release point counting down from the five kilometre beam. This was much more accurate than the *Knickebein*, allowing the fall of bombs within 300 metres from a bomber at 20,000 feet.[4] It was this specialised pathfinding role by KGr100 that led to the devastating raid on Coventry. However, the British were able to effectively jam *X-Gerät* by late 1940. A technique used to counter this system was to insert a false cross-beam, short of the planned point of intersection, which triggered an early bomb release.

This electronic activity not only degraded *Luftwaffe* operations by causing bombs to fall wide of the target, negating to some extent the German offensive, but the signals from enemy navigation beacons also leant themselves to interference. The signals from German beacons picked up in England, were retransmitted with sufficient strength to mask the original tones. This masking of beacons -"meaconing" - misled German aircrew into obtaining erroneous navigation fixes, and became part of the "battle of the beams", which was to develop into what was to become electronic warfare.

With the failure of *X-Gerät*, the Germans countered at the end of 1940 with the *Y-Gerät* system.[5] This operated on one beam only, a single beam from a ground station transmitted on 42.5 MHz, which was re-radiated to the station by an aircraft transponder on 46.9 MHz. The ground station therefore knew the position of the aircraft along the beam, and transmitted a bomb release signal in the target area. Once again the British had devised a counter. On receipt of the aircraft transponder signal, the British jammer would transmit on 42.5 MHz, which would be received by the bomber and again trigger its transponder. This system of triggering extra transmissions gave the ground station in France a false track distance. After this successful jamming in January 1941, the use of *Y-Gerät* equipment was soon abandoned.

These transmitted signals had been a windfall for the British defences. The Germans were providing a warning system of the direction of attack, making it possible for the alerted defences to have fighters or anti-aircraft fire in the right place at the right time. These systems rendered Britain far greater service than they did Germany, and had been made possible by the British breaking the code of the German Enigma cryptographic machines, the closely-guarded secret known as Ultra.

Confusing the enemy was, however, only one part of solving the problem. For any long term resolution, the enemy had to be found and destroyed. Radar, still in its infancy, performed poorly over land. One method used to locate enemy aircraft, to guide the searchlights and anti-aircraft guns, was sound detection. However, the performance of sound-

4 *X-Gerät* worked on 4 metre wavelength, the director beam frequency being 66.5 to 71.5 MHz, and the cross beams 71.5 to 75 MHz, with a maximum range of 250 miles. Jones, op cit, p. 203.

5 *Y-Gerät* worked on 7 metre wavelength

The Defiant, above, and the Blenheim were day fighters that had been relegated to the "cat's-eyes" night fighter role because with their poor performance, they could not survive by day.

locaters varied greatly with the weather, and their effectiveness had been outstripped by the high speed of modern fighters. An aircraft flying at 300 mph at 20,000 feet would be three kilometres further on before its sound reached the listening apparatus on the ground, and ten kilometres further on before the shell bursts of the anti-aircraft fire.[6] In addition, there were further limitations. Searchlights could not penetrate cloud, could not track aircraft for a sufficient length of time, nor illuminate effectively above 12,000 feet. Moreover, the anti-aircraft gun defences were inadequate in both numbers and performance.

Night fighters had been almost totally neglected. Radio communication between aircraft and the ground was often indecipherable, and restricted to short ranges. Blind-flying instrumentation was poor, airfield lighting was insufficient and the grass runways while suitable for day operations, were hazardous for any movements at night. These grass strips were dimly lit with primitive flood lights and paraffin flares, with no system of homing or provision for blind approaches.[7] Once airborne, it was difficult to navigate

6 Richards, op cit, p.201.

7 Soon the Lorenz Standard Beam Approach would be introduced.

without radio beacons, and to find an elusive target on a dark night was an ineffectual chase around the sky. One step taken was the equipping of night fighters with Lorenz blind-flying equipment, which homed on the *Knickebein* signals, enabling the fighters to hunt up and down the beams.[8] But this was providing only part of the solution, and did not result in great success. Not only did a night fighter need information of its target in azimuth, but also in elevation and in range. The answer lay in radar.

Britain forged ahead in the radar race, driven by this requirement for the detection and location of German bombers at night which needed to be destroyed. The urgent task required radar equipment light enough to be carried by aircraft. The aircraft initially selected by the RAF as its night fighter was the Bristol Blenheim. Designed as a twin-engined bomber, the Blenheim If was adapted to the "cat's-eyes" night fighter role when No 25 Squadron equipped with the type in 1938. Over 1939 and 1940, Nos 23, 29 and 600 Squadrons also joined 25 Squadron equipped with night fighter Blenheims, and the Fighter Intercept Unit began trials with Blenheims equipped with Airborne Interception (AI) radar.

British scientists then had found that the shorter the wavelength (ie the higher the frequency), the clearer the image on the radar screen. However, with this higher frequency AI radar, the RAF took the wise precaution of not permitting aircraft to overfly enemy territory. The Germans had been working on decimeter wavelength radar, regarding the centimetre wavelengths as impracticable. Early British AI radar operated on 1.5 metre wavelength, and could detect aircraft at a maximum range of between 3 to 4 miles.[9] On the night of Monday 22 July 1940, a Dornier Do 17 (of 2/KG3[10]) was shot down by an AI Mark III radar-equipped Blenheim of the Fighter Interception Unit. This was the first 'kill' for a radar night fighter.[11]

At the beginning of the war, the Germans had not thought as much about the use of radar as the British. The reverse had been the case with regard to the navigation beams for the bombers. The British had been thinking defensively, the Germans more of offensive action. The Germans failed to appreciate the highly effective system of fighter control integrated with the radar detection chain of stations, and virtually ended its radar research. Efforts were directed instead towards increasing aircraft production, and German research on centimetre wave radar ceased.[12]

A problem for AI-equipped fighters was that the fighter had to be within 3 to 4 miles of the target aircraft to detect it. This problem was solved too by British scientists. They developed a radar for long-distance inland tracking over the summer of 1940, and this had allowed the first station to become operational at Durrington, near Shoreham, West Sussex, on 16

8 Richards, op cit, p.209.

9 K Bartz, *Swastika in the Air*, Kimber, London, 1956, p.55.

10 2nd Staffel (squadron) of KG3, Kampfgeschwader 3 (3rd bomber wing).

11 K Heeks, *Bombing 1939-45*, Hale, London, 1990, p.98, and Richards, op cit, p.203.

12 Bartz, op cit, p.56.

October 1940. This Ground Controlled Intercept (GCI) radar operated on 1.5 metre wavelength, with a rotating antenna, which revolved once every twenty seconds (ie 3 RPM). The return from the target was displayed on an opaque glass screen, the Plan Position Indicator (PPI), as a bright spot, providing bearing and distance (ie position) from the radar site. Height, speed and direction of the target could also be measured out to a range of 70 kilometres. The first GCI was ready because it had been given, at Churchill's intervention, the highest priority in resources.

The GCI operator was now able to observe the enemy aircraft, and control the intercepting fighter by calling directions over the radio. Night fighter crews were to work intimately with the ground controller. The GCI controller called in a night fighter from its patrol line, and by directing the pilot to fly appropriate headings, he juggled the two "blips", or radar

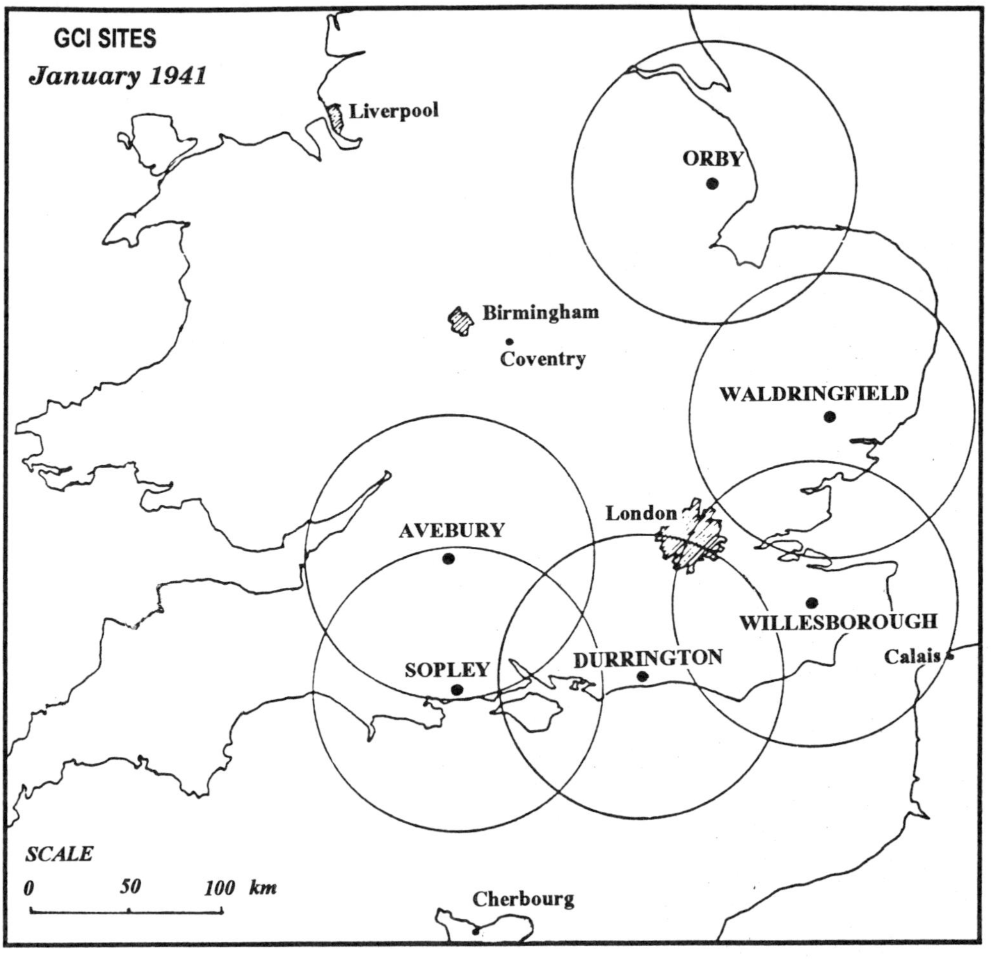

After the first GCI became operational at Durrington, it was followed by another nearby at Sopley, near Christchurch, Hampshire, and by January 1941, there were six GCI installations operational covering England's south-eastern coastline.

returns, on his scope, until ideally the night fighter was manoeuvred to three miles behind the bomber. The GCI controller would then instruct the night fighter to "flash your weapon", thereby enabling the fighter to take over the attack by means of the AI radar. After the intercept, the GCI controller sent the night fighter after another target, or handed it back to the sector controller for homing to its airfield. Initially it had been pointless sending up more fighters than the controller could handle, but as GCI operators gained more proficiency, they were able to control multiple engagements simultaneously. Like AI, GCI too was to play a decisive role in night fighting.

From the beginning of November 1940 until the end of February 1941, the enemy had launched a further 12,000 sorties over the UK. The defences had destroyed some 75, with approximately two-thirds being destroyed by guns, and one-third by the night fighters.[13] By March 1941, five of the six AI-equipped squadrons of Fighter Command had exchanged their interim night fighter Blenheims for new Beaufighters. The following month, there were eleven GCI stations in operations to guide them. With the introduction of GCI control from January 1941, night fighter kills mounted rapidly. On the heavy blitz on London on the night of 19/20 May, 24 of the enemy were shot down by fighters compared with only two by AA guns.[14] The Blitz was virtually ended, and over the summer of 1941, the incidence of German raids dropped sharply, which was to lead to a period of reduced activity for the night fighters of Fighter Command.

The improved night fighter effectiveness over the first half of 1941 was due to the introduction of the Bristol Beaufighter with AI Mark IV radar and the efficient use of GCI, as the following table shows:[15]

Month	**Jan**	**Feb**	**Mar**	**Apr**	**May**
Guns	12	8	17	39	31
Night Fighters	3	4	22	48	96

However, there were insufficient Beaufighters to equip the new Australian night fighter unit, No 456 Squadron, which was forming in June 1941. This new RAAF squadron received instead the Boulton Paul Defiant, a single-engined fighter which had been adapted to the night role without AI. These were known as "cat's-eye" fighters, whose pilots relied on vectors supplied by sector operations towards the plots of enemy raiders. To find their prey, these crews stumbled around in the dark, relying solely on their eyesight. It

13 Richards, op cit, p.213.

14 P J R Moyes, *Bristol Beaufighter I & II*, Profile Publication No.137, Leatherhead, Surrey, p.10.

15 Richards, op cit, pp.214-5.

was rumoured that "cat's-eye" crews were fed on a staple diet of carrots! For these night fighters, hunting enemy bombers was to become a hopeless task.

CHAPTER THREE

THE DEFIANT

As part of the process of forming Australian Article XV squadrons under the EATS agreement, aircrew were gathered at different bases in England to begin their operational work-ups. Concurrent with this, groups of Australian groundcrew were being organised into a variety of units at the RAAF base at Williamtown, near Newcastle, New South Wales, and among these a group would form the nucleus of ground staff for 456 Squadron.

The aircrew assembled for the new squadron at Fighter Command's 9 Group base at Valley, on the windswept island of Anglesey, in northern Wales, from 30 June 1941. Their CO was an Australian pilot serving with the RAF, Squadron Leader Gordon Olive DFC[1]. Olive had trained at Point Cook in 1936, before joining the RAF on a short service commission early the following year. He had flown Gloster Gauntlet and Gladiator biplane fighters with No 65 Squadron, becoming a long-serving member of this unit, and had been elevated to 'A' Flight Commander on Spitfires during 1940. After operations over France, he flew from Hornchurch during the Battle of Britain, destroying four German fighters, and claiming a further six as probables. This resulted in his award of the DFC during September 1940. He destroyed another fighter during December 1940, which had entitled him to be classified as an "ace".

During the Battle of Britain, Gordon Olive had one of those days when nothing seemed to go right. He was leading the Spitfires of No 65 Squadron taking-off from Hornchurch on the afternoon of Wednesday 7 August, when

1 Wg Cdr C G C Olive MBE DFC, 39469 RAF/277457/O17934. No 65 Sqn RAF, ftr control Tangmere, 219 Sqn; comd 456 Sqn 1941-42; Air Def HQ Sydney, Air Def HQ Morotai. Later Qld Sqn ATC. Regular air force offr of Brisbane; b Brisbane 3 Jul 1916.

his oxygen regulator caught on fire, which filled his cockpit with smoke. Opening the canopy to clear the smoke, he realised he had to escape from the imminent explosion. But, he was trapped in the cockpit:

> He cannot roll the plane over and fall out because he does not have enough height for his parachute to open. Also, with 11 Spitfires in formation behind and below he could be struck by someone's churning propeller. He must do something...and quickly. Heat is searing his face.
>
> Olive's only chance is to try a stall turn to gain some height quickly and clear the others. Waving his wingmen away, he pulls back into a vertical climb, undoing his harness at the same time. While zooming upwards he pulls off his flying helmet and oxygen mask and then pushes open the entry hatch. He has gained some 500 – 1,000 feet (150 – 300m), still not a great margin for error but there are no choices left. Standing the Spitfire on one wing, he kicks himself out of the cockpit and pulls his ripcord...but nothing happens. The parachute has not opened. Looking down, he sees that the small pilot chute has caught in his flying boots. While hurtling through the air he manages to untangle it and push it away so it can drag out the main canopy. A shoulder-wrenching jolt almost knocks him unconscious as the parachute opens and jerks him upright.
>
> It takes a few seconds for Olive to recover his senses and realise that he is now floating gently down. He then notices something which terrifies him...He is drifting towards high tension cables.
>
> He must try to side-slip out of their way but, as he reaches up to manipulate the cords of his parachute, he gets another shock...The

Wing Commander Gordon Olive DFC, first Commanding Officer of No 456 Squadron

canopy is splitting in two! Any attempt to side-slip might collapse the chute completely...

As if these problems were not enough, the quiet is suddenly shattered by two loud reports. Someone is firing at him! Looking down again, he spots two home guardsmen who are shooting upwards with shotguns, but fortunately they are still out of range.

Everything seems to accelerate. He is dropping quickly now but at least he is going to miss the cables. More shots go wide. He thuds down into a potato field and the softness of its soil breaks his fall. He is uninjured and safe...or is he?

The two home guardsmen now rush up and demand to take him into custody. Olive tells them that he "is not going to bloody well go", and finally two apologetic guardsmen retreat from the scene.

Seconds later, they are all surprised by the arrival of a fire engine and an ambulance. The whole incident had been witnessed and they have charged across country to the rescue. Olive agrees to ride back in the ambulance and waits while the driver turns around...but, unfortunately, it strikes a ditch and topples over. Luckily, the driver scrambles out uninjured.

The Australian now decides to return on the fire engine which will take him to Hornchurch. However it arrives at an unexpected hairpin bend and fails to negotiate the turn. It crashes through a rail and down into the water. Olive manages to jump clear and he last sees the engine as it disappears in foam and spray. Luckily, except for wounded pride, nobody is hurt.

After climbing back to the road, Olive tucks his parachute under his arm and begins to walk home. There cannot be too many of his nine lives left!

An old bull-nosed Morris utility pulls up beside him and the driver calls out:

"Hey, can I give you a lift?"

"Not bloody likely", Olive replies, "I'm going to walk!"[2]

Olive's job was to mould the new unit into an effective fighting unit, as the RAAF's first night fighter squadron. No 456 Squadron, like other Australian units formed under EATS, was RAAF by name, intended to have a large Australian complement of air and ground personnel, but was controlled by the RAF, under RAF command, and equipped with RAF aircraft.

Meanwhile, the ground personnel for the new EATS squadrons were being assembled at the RAAF base at Williamtown, over May and June 1941. No 454 Squadron had formed during May, but plans had changed and it was soon disbanded, to form again later. The personnel assembled for this unit were either absorbed into another two new squadrons, Nos 455 and 456, or held at Williamtown to form No 458 Squadron. One of the airmen posted to the new 456 Squadron was LAC Len Stamp[3]. He recalled:

2 D Newton, *A Few of The Few*, AWM, Canberra, 1990, pp.79-80.

3 Sgt L Stamp, 23533. 1 BAGS, 3 SFTS, 456 Sqn, 6451 SE. Apprentice watchmaker of Brisbane; b Ipswich, Qld, 10 May 1922.

> At Williamtown, being issued with tropical gear, we were rumoured to be for Singapore. In June 1941, we were paraded and handed in tropical gear for blue uniforms and heavy gear. We then became 456 Squadron, 455 Squadron being formed at the same time, it then being apparent that we were to travel together – wherever![4]

Known as the "Blue Draft", the airmen of both squadrons departed Williamtown the night of 16 July for Sydney, under the command of Flight Lieutenant John Pike[5]. As an elder and more sober member, he soon was known as "Granny" to the young airman, and was to become 456's first adjutant in the UK. With the members of No 455 Squadron, which was forming in UK as an RAAF squadron for Bomber Command, the 182 airmen and two officers of 456 Squadron's ground staff embarked the following day aboard the SS *Johan Olden Barnveldt*.[6] Being picked up outside Sydney Heads by an escort, HMAS *Adelaide*, the *Johan* – "a dirty stinking scow of a thing"[7] – made the crossing to Auckland in four days.

The Australians changed ship at Auckland, transferring to the HMT *Dominion Monarch*, which after the previous ship, was considered a palace. Departing New Zealand on 23 July, the *Dominion Monarch* zig-zagged south, and then eastwards towards South America, to then follow the coast up to Panama. Passing through the Panama Canal on 7 August: "One could only be impressed with the engineering and construction of the Canal, understanding the strategy of the US involvement and policing of it."[8] Passing through Curacao, the *Dominion Monarch* was then escorted by the Royal Canadian Navy through the West Indies and up the US east coast to Canada. For a week the Australians were granted day shore leave in Halifax, Nova Scotia, before departing across the grey, uninviting rolling Atlantic for England. Arriving in Liverpool late in the evening of 1 September, the passengers were "appalled at the shipping destroyed around the approaches and up the Mersey – even before tying up, the barrage balloons were up over us".[9]

Disembarking next morning, the Blue Draft entrained for the personnel processing centre at Bournemouth, on England's south coast.[10] After initial processing, 180 men joined the new Australian night fighter squadron at Valley, on the Welsh island of Anglesey, to join the predominantly RAF aircrew working up as No 456 Squadron.[11] However, with both the CO and

---

4 Len Stamp, letter to author, 11 Dec 1994.

5 Sqn Ldr J E Pike, 271018. Nos 4, 454 & 456 Sqns, 27 OTU. Company director of Brisbane; b 19 Dec 1912.

6 No 456 Sqn A50, Operations Record Book, 17 Jul 1941.

7 Len Stamp, letter to author, 11 Dec 1994.

8 ibid.

9 ibid.

10 Jack Benning, interview with author, 7 Jul 1994.

11 J Herington, *Air War Against Germany and Italy 1939-1943*, RAAF Official History Vol III, AWM, Canberra, 1954, p.145.

HMT Dominion Monarch

the Adjutant hailing from Brisbane, it was not long before 456 was being referred to unofficially as the "City of Brisbane Squadron".[12]

Valley was a new aerodrome, built on sand dunes, with the airfield facilities fully dispersed. The domestic accommodation was dispersed also, with the officers and airmen alike detached to various numbered sites in the surrounding villages. The Australians soon became familiar with the bicycle as a means of travel. Eventually, the airmen were all brought together in barracks at Site 8.

As only the very first graduates of the EATS were starting to reach squadrons, initially Gordon Olive's aircrew members were pilots and air gunners of the RAF, including his 'A' Flight Commander, Flight Lieutenant John Hamilton[13], and 'B' Flight Commander, Flight Lieutenant C Cooke[14]. The first RAAF aircrew joined the Squadron during July, including pilots Pilot Officer Tom Cleary[15], with Sergeants Gwynne "Big" Hughes[16] and Jack Kirkman[17]. The air gunners soon followed – again early EATS graduates –

12 Officially, the City of Brisbane Squadron was No 23 Sqn RAAF.

13 Wg Cdr J S Hamilton DFC AFC RAF, 73752. No 256 Sqn RAF, 456 Sqn, 23 Sqn RAF, 418 Sqn RCAF; comd 613 Sqn RAF 1944, Wg Ldr 138 Wg 1944-45. Regular air force offr of Portsmouth, Eng; b Portsmouth 1 Sep 1911.

14 Sqn Ldr C A Cooke DFC RAF, 43634. No 456 Sqn; comd 264 Sqn 1941-42. Regular air force offr; b 7 Jun 1912.

15 Flg Off T Cleary, 404236. Nos 456 & 30 Sqns. Pharmacist of Lismore, NSW; b Byrnestown, Qld, 3 Aug 1916.

16 Flt Lt E G Hughes, 407139. Nos 456 & 30 Sqns, 1AD, 1 WAGS, 6 SFTS. Stock clerk of Adelaide; b Adelaide 30 Oct 1916.

17 Flt Lt J M Kirkman, 406017. No 456 Sqn, 285, 232, 64, 229 & 603 Sqns RAF. Public servant of Perth; b Northam, WA, 12 Jun 1918.

including Pilot Officer Dan Norris-Smith[18], destined to become a long-term 456 member, and Sergeants Tom "Little" Hughes[19] and Ron Lowther[20].

The only equipment available for 456 at Valley was the hopelessly outdated and outclassed Boulton Paul Defiant. The Defiant was a single-engined fighter with a crew of two – pilot and air gunner. By no stretch of the imagination could it be considered a night fighter. It was not equipped with navigational aids or with radar for the night role. To find enemy bombers during the Blitz, the Defiant had groped around in the dark, with crews looking out of their cockpits for any raiders against the glow of the fires.

Defiants had equipped No 264 Squadron RAF during May 1940, when, operating from Manston as an advanced landing ground in Kent, they had flown patrols over the beaches at Dunkirk and Boulogne. The rear gun turret of the Defiant meant that a stern attack onto the aircraft could prove fatal. Initially, the Defiants were misidentified by the Germans as single-seat fighters, and on one occasion on 13 May, four Junkers Ju 87 Stuka dive-bombers and a Messerschmitt Bf 109 fighter were destroyed by the rearward-firing guns. A wartime aviation magazine stated: "When the Boulton Paul Defiant first went into operational service it carried with it a useful but short-lived weapon of war – that of surprise".[21] However, once Luftwaffe pilots could identify the Defiant from the Hurricane, the Defiants of 264 Squadron became "easy meat", and towards the end of May suffered too high a loss rate.[22]

During July 1940, the Defiants of No 11 Group, began Channel patrols. During one fierce battle on Friday 19 July, nine No 141 Squadron Defiants had departed Hawkinge, in Kent, to patrol south of Folkestone. After a mauling by Bf 109s, only four limped back to the airfield – one then crashed near Hawkinge village, and another belly landed. Four pilots and six gunners had been killed in the heaviest defeat suffered by a Fighter Command unit, and the episode became known as "the Slaughter of the Innocents".[23] The unit immediately returned north to Grangemouth, in Scotland, for convoy patrols, and in September began night fighter training. The following month, after further decimation of Defiants, it was at last obvious that the aircraft could never again be risked in daylight operations.

Over the summer of 1941, the Defiants had operated with the "Turbinlite" Havocs of Fighter Command. The Turbinlite modification was

18 Gp Capt D L Norris-Smith OBE, 400309/later RAF. No 256 & 23 Sqns RAF, 456 Sqn, HQ ADGB, 68 Sqn RAF, 5 OTU. Comd 23 Sqn 1947-49; HQ NEAF. Station overseer of Deniliquin, NSW; b Adelaide 9 Oct 1916.

19 Flt Lt T C Hughes, 408086. No 456 Sqn, 275 & 277 Sqns RAF. Grocer of Launceston, Tas; b Launceston 20 Jul 1915.

20 Plt Off R C Lowther, 404179. No 456 Sqn. Company representative of Brisbane; b Brisbane 1 Jan 1916. Killed in action 2 Jun 1943.

21 "Their Characteristics – Boulton Paul Defiant", in *Flight*, London, 25 Feb 1943.

22 Rawlings, op cit, p.378.

23 Newton, op cit, p.52.

another experiment in developing a credible night fighter deterrence. The Douglas Havoc aircraft would find a target by use of its airborne intercept radar, and then illuminate its nose-mounted Turbinlite searchlight to allow an accompanying fighter to shoot down the target. It proved an unwieldy operation and was never a successful tactic. In December 1941, No 151 Squadron was to develop a "Flare Burning" tactic, whereby a Defiant would fly with a Havoc, to pick up any low-flying enemy aircraft with its AI and drop flares on it, enabling the Defiant to attack. Again, as with the airborne searchlight, 151 Squadron never scored any victories with this method.

Training for 456 Squadron in these formative months over the second half of 1941 involved aircraft conversions onto the Defiant during the day, and operational role training during the night. The facilities at Valley were considered among the best in the country, with sealed runways, 4,500 feet to 6,000 feet long, with good approaches. Placed on the very edge of the sea, it was quite often open while most others were closed by weather. However, the Australians soon found to their chagrin that the base was subject to very strong winds. As winter approached, in addition to being windy, it was normally raining heavily and very cold. One serious disadvantage of the aerodrome was that the runways were set among numerous sand dunes, and pilots under instruction found they could not swing off the runway with impunity.

For the operational night fighter role, Valley appeared conveniently situated for the defence of the industrial areas of Lancashire, particularly from raiders coming up the Irish Sea. However, with no aids to assist the task of night interception, the Defiants often appeared to be chasing about in the dark. The crews constantly ran the risk of being coned by their own searchlights, then accompanied by the resultant anti-aircraft fire. The Valley sector was to prove a backwater area, with very little "trade" coming that way.

The Defiant's disastrous losses relegated the ttype to night fighting

A Defiant in the dispersals at Valley, 1941

The training program was soon marred by tragedy when, on 21 July, one of the RAF pilots, Sergeant A Brookes[24], was killed when flying solo in a Defiant (T3933). He crashed at Llanerchymedd, north of Valley, in the centre of the island. Despite this loss, the first air fatality for the new Squadron, the flying safety record during this intense period of training was to remain good. What was not good, however, was how ineffective the Defiant was for night fighting. Fortunately, the Defiant was not destined to remain long with 456 Squadron. The first indication of new equipment for the unit was in early August, when the first 456 crew was attached to 68 Squadron at RAF High Ercall, in Shropshire, for Beaufighter familiarisation.

Meanwhile, 456 was achieving between 150 and 200 hours flying effort on Defiants weekly, and of the Squadron's 18 pilots at the end of August, eleven had been declared operational. The Squadron had now settled into the normal routine of a night fighter unit, with one Flight on duty for two nights, rotating then with the other Flight for two nights off. It was a sacred rule on the squadrons, that before flying operations at night, each crew on the flying program would fly in the afternoon on a night flying test (NFT) to check the aircraft before flying their patrols.

A close bond was soon established between the Australian aircrew and ground staff. Whether this was engendered because they were the first of the EATS graduates and groundcrew drafts, or whether they were in the isolation of Valley, a long way from home, is difficult to fathom. But the

24 Sgt A F Brookes RAF, 1166497. No 456 Sqn. Killed in flying accident 21 Jul 1941.

behaviour between the flyers and the groundcrew appeared unique to the British members working beside them. There was nothing of the normal formality of the RAF squadrons, where the casual attitudes of the Australian mechanics most probably would not have been tolerated. It was that Aussie egalitarian style, not one of disrespect, but one of buckling down to get the job done, in a casual fashion. Each of the two Flights were directed by a Flight Sergeant: 'A' Flight's was Johnny Goodall[25], and 'B' came under Keith Duncan[26]. Flight Sergeants Eric "Ali" Barber[27] and Lance Sutherland[28] would also later figure prominently on the ground operation. One pilot described the management style:

> There was less emphasis on rank in 456; we were all on the same team with the same goal to achieve, so lived and worked pretty much at the same "level". The ground crews were terrific, working endless hours, day and night in all sorts of weather. We never had any cause to feel uneasy when flying. Things worked and kept on working.[29]

One of 456's air gunners was to win an award for gallantry over this period, but it was not as a result of operational flying. When a Blackburn Botha reconnaissance aircraft crashed into a very heavy sea off Valley during August, many people on the shore attempted to swim out to rescue the three crew. Sergeant John Plunkett[30], who had been a life-saver at Cairns before the war, swam out 300 yards through the raging sea and brought back an unconscious gunner, whom he attempted to revive by artificial respiration. Many of the would-be rescuers were also caught by the strong sea and were drowned. Sergeant Plunkett was able to bring three others to shore, and these owed their lives to his courage.[31]

John Plunkett was to lead an interesting war. Remaining as an air gunner after 456 re-equipped with Beaufighters, he was posted to No 458 Squadron on Wellington bombers, and soon transferred with that unit to the Middle East. 458's aircrew were then dispersed to several RAF Wellington units during 1942, and while flying an attack on Tobruk on 7 August with 104 Squadron, he was posted as missing on operations. The Wellington had

25 FSgt J H Goodall BEM, 5355. Nos 25 & 456 Sqns, 3064 SE, 6456 SE, 4466 SE. Assistant engineer of Adelaide; b Adelaide 5 Nov 1920.

26 FSgt K J Duncan, 23191. 3SFTS, Nos 454 & 456 Sqns, 3064 SE, 6456 SE. Engineer of Byron Bay, NSW; b Clunes, NSW, 28 Feb 1904.

27 FSgt E Barber BEM, 5300. No 456 Sqn, 3064 SE, 6456 SE, 1 AD, 1 APU. Motor fitter of Melbourne; b Croydon, Surrey, Eng, 20 Dec 1917.

28 FSgt L M Sutherland, 6944. Nos 455, 452 & 456 Sqns, 3064 SE, 6456 SE, 4466 SE. Motor mechanic of Murwillumbah, NSW; b Lismore, NSW, 20 Jun 1920.

29 Fred Stevens, *An Aussie in a Mossie*, unpublished manuscript, 1991, p.34.

30 WOff J J Plunkett BEM, 404216. Nos 456 & 458 Sqns, 104 Sqn RAF. Clerk of Buninyong, Qld; b Cairns, Qld, 21 Jun 1912. POW 8 Aug 1942.

31 London Gazette 35404 of 6 Jan 1942, Overseas HQ P.1/43 of 1 Jan 1943. RAAF Historical Section, Canberra.

belly-landed in the desert, and the crew, although escaping largely unscathed, were captured. Plunkett was to see out the rest of the war as a POW, mainly at Stalag VIIIB in Lamsdorf, Germany, but his war was far from over. He exchanged identities with another prisoner, Corporal E Smith, and under his assumed identity was able to gather intelligence information, which he sent back to England by postcards. He identified targets suitable for attack by "little stinging insects" (Mosquitos) or by "our lanky friends" (Lancasters). Plunkett reverted to his real identity, and was released at the war's end.

Two more aircraft accidents occurred on 456 Squadron during early September. While night training in the early hours of 3 September, Sergeant Jack Kirkman had an undercarriage oleo leg break off at the end of his landing run at Valley, which caused minor damage to his Defiant N3458. The following day, Pilot Officer Tom Cleary forced landed at Squires Gate, near Blackpool. As he made his approach, the aircraft struck some trees, resulting in the write-off of his Defiant, V1177.

In spite of incidents like these, No 456 Squadron was now ready for battle. The first operation was carried out on 4 September, as two Defiants flew convoy patrol. The only action was when the convoy opened fire on Flight Lieutenant Cooke's aircraft (N3497), fortunately with inaccurate results. Uneventful operations were continued with Defiants throughout September, and the first Beaufighter (R2474) arrived on 26 September, followed by a further five by the end of the month.

The changeover from the Defiant to the twin-engined Beaufighter had the same affect on 456 Squadron, as it had on other night squadrons of Fighter Command. Some pilots preferred single-engined aircraft and did not want to convert to "twins". For the air gunners, who had squeezed into the turret of the Defiants, it was indeed a completely different trade. The back seat of the Beaufighter required a new breed of navigator, known as an Observer Radio, abbreviated to RO, who was basically a radar operator. The first ROs to reach the night fighter squadrons had been technicians who had worked in France before the evacuation on the early radar ground chain. Some gunners were not willing to become ROs, and were posted away to other units. Those who were keen to change left Valley for No 51 Operational Training Unit (OTU) at Cranfield, near Bedford, for ten days of training as radar operators.[32] Later, as training became more structured, radar training OTUs (such as 62 OTU RAF Usworth, in the north near Newcastle-upon-Tyne) were formed to teach ROs this new profession, in Anson flying classrooms configured for radar instruction.[33] Then ROs would normally team up with their pilots at the night fighter OTU (or in some cases, on their squadrons), to begin developing their close-knit teamwork.

32 Eventually, on 23 July 1942, ROs were recategorised as Navigators (Radio), sometimes abbreviated to nav/rads.

33 AI training had initially been conducted at the Radio School at Prestwick, near Ayr, before moving to 62 OTU at Usworth.

456 Squadron 'B' Flight crewroom at Valley in late 1941, with the visit of representatives of the Bristol company and the Australian press, after the announcement of the Squadron's re-equipment with the Bristol Beaufighter. Wing Commander Gordon Olive DFC is in the front centre. Aircrew at rear, from left, are Flying Officer Reg Pargeter, Pilot Officer Meredith, Flying Officer John Fell and Sergeant Ray Hedge.

During October, with the arrival of the new aircraft, Flight Lieutenant "Dot" Dottridge[34] was posted in to lead 'B' Flight:

> He was a permanent RAF officer, looking older than his 26 years, with a shock of grey hair and a magnificent handlebar moustache. His log book warranted more than a cursory glance, having flown Blenheims in daylight over France and Germany until that became slightly suicidal.[35]

The last Defiant operation was flown on 20 October, and although intercepts were attempted on several raids, no contacts were obtained. This night role had been a task the Defiant had assumed by default, and night fighting in the Defiant had truly been a hit or miss affair. The Defiant was not up to the demanding role of night operations, but its three months' service with 456 Squadron was that of advanced training for the new unit, and provided an insight and grounding for the novices, of the difficulties they would confront in the night war.

34 Gp Capt A P Dottridge DFC* RAF, 41907. 456 Sqn, 54 OTU; comd 68 Sqn RAF 1943, 256 Sqn 1946. Regular air force offr; b 8 Sep 1916.

35 Stevens, op cit, p.26.

'B' Flight Valley at the beginning of 1942:
Rear, from left: FSgt Wally Cox, Sgt "Red" Ratcliffe, Sgt George Postlethwaite, FSgt Bill Williams DFM, Sgts Tommy "Little" Hughes and Gwynne "Big" Hughes, Sqn Ldr "Dot" Dottridge DFC ('B' Flight Commander), Flying Officers Reg Pargeter and John Fell (a crew that later became "aces" with 29 Sqn RAF and who were both awarded DFCs), Sgt Ray Hedge.
Front: Sgts Cliff Satchwell, "Spike" Burley and Tom Condon, Plt Off Tom Cleary, "Leopard" the dog, Sgt Jim Telford, Plt Off Bernie Wills.

The mode of transport at the dispersed base of Valley was bicycle. From left: Don Shanks, unknown, Roger Hitchcock, Fred Stevens, Geoff Oxlade, John Brayne, Red Ratcliffe, Ron Lowther, Tony Samson, Dickie Williams, Cliff Satchwell, Mark Stokes and Alec Abbey.

CHAPTER FOUR

THE SECRET OF THE NIGHT

The next British ground radar development from the Chain Home (CH) early warning system had been the adoption of a Ground Control Intercept (GCI) system. Interception had initially been done by the sector operations rooms, from filtered information of the plots provided by the CH and Chain Home Low (CHL) radar stations, looking out to sea, and the reports of the Observer Corps, as the raid progressed over land. Controllers passed instructions to the fighters by radio through the normal raid-reporting chain, but the results were usually inaccurate. GCI provided an accurate overland radar capability, with a ground operator plotting fighter and quarry on a 360 degree scan radar scope, the Plan Position Indicator (PPI). The ground controller was able to conduct accurate intercepts by tracking the enemy aircraft, and vectoring the defending fighter into a tail chasing engagement. By July 1941, Fighter Command's defences had grown to 17 operational GCI stations, feeding 16 squadrons of night fighters.[1]

GCI radars made it possible for fighters to be accurately directed onto German raiders, and the ground controllers could vector the fighter to within three miles of a target. What was now required was to close that last three mile gap to the target. This was the role of airborne interception (AI) radar, which the British were able to retain as the very well-kept secret of the night. Initially, this radar research and development was in the metric wavelengths, in the VHF radio band.

1 Bowyer, op cit, p.94.

AI Mark I

In 1937 an experimental airborne transmitter, known as RDF-2, had been trialled in an Avro Anson, and by the following year airborne interception sets were contracted for production. The intent was to fit AI sets to aircraft with two-man crews.

The AI Mk I first flew on 21 May 1939 in a single-engined Fairey Battle.[2] The results were generally unsatisfactory, and on 10 January 1940, Headquarters Fighter Command wrote that AI Mk I was of no operational value.

AI Mark II

The AI Mk II was the productionised version of the Mk I, weighing about 270 kg (600 lb). Modifications were introduced to increase the power output and reduce the minimum range. For AI to be of any use at all, the minimum range had to be reduced to a range where an aircraft could be discerned by the naked eye at night. If the equipment could achieve this, then the target could be identified and attacked.

A production run of 21 sets was completed during 1939, and installed in the twin-engined Bristol Blenheim If, which was then equipping the night fighter squadrons. The first three were in service with No 25 Squadron just prior to the outbreak of war, and by December 1939, all sets had been fitted to Blenheims. The Fighter Interception Unit (FIU) was formed on 12 April 1940 to test AI radar and develop night fighting techniques. One of the first tasks of FIU was to test AI Mark II, which like its predecessor, also proved unsatisfactory.

AI Mark III

The Mk III, with an improved aerial system to alleviate ambiguity problems, first flew in August 1939. On 3 May 1940, the decision was made to fit AI Mk III, on the highest priority, to 100 Blenheim aircraft, and in three months Fighter Command was able to report AI was fitted to 69 night fighter aircraft.

The maximum target range was limited to the ground reflections to the height of the AI-equipped fighter - allowing an operationally useable range window of about 1.5 to 2 miles, to detect and visually identify the target, and bring the target within guns range.[3] Although this may have been possible by skilled operators in trials, operationally the task proved impossible for the AI Mk III equipped Blenheim crews.

In fact, despite the success mentioned previously of a Blenheim in the downing of the Dornier Do 17 over the Channel on the night of 22 July 1940, virtually the only other contacts made with AI Mks I, II and III were on practice interceptions. In the face of growing scepticism, this primitive

2 B Gunston, *Night Fighters*, Patrick Stephens Ltd, Cambridge, 1976, p.47.

3 D Martin, *50 Years in Radar*, Thorn EMI, Hayes, 1986, p.4.

form of AI was doubted as a useful aid.[4] The AI Mk III set had demonstrated a minimum range of between 1000 and 1500 feet, while the effective lethal range of the aircraft's Browning 0.303-inch guns was about 600 feet, leaving up to 1000 feet of tracking to be covered visually, usually by tracking exhaust flames. This was soon shown to be operationally impossible to achieve, particularly as simple flame shields were fitted to all German bombers.

Trials held in February 1940 showed that a development of the Mk III set produced a much improved minimum range, which not only virtually guaranteed visual contact, but also allowed accurate target range to be fed to the interceptor's guns up to the point of firing. This became the AI Mk IV, the principal AI radar used by the RAF (and also by early US night fighters) during the war.

AI Mark IV

The AI Mk IV, known officially in the RAF as ARI 5003, was first tested by FIU during the night of 13/14 August 1940, and opened new possibilities by its three to four mile range, with a minimum range of only 600 feet.[5] In addition, AI Mk IV had a wider angle of scan than the Mark III, enabling a 40 degree cone ahead of the fighter to be searched, and had a tracking accuracy of 10 degrees.[6] Its arrival coincided with the introduction of the twin-engined Beaufighter, which offered an ideal platform with its two-man crew. AI Mk IV became the standard fit for Fighter Command from September 1940, and 600 sets were ordered, to be produced by Pye, EMI and GEC.

The first Beaufighter kill came on 19 November 1940, when Squadron Leader "Cats Eyes" Cunningham shot down a Junkers Ju 88 over Norfolk. But although an excellent night fighter with an efficient AI had just entered service, night defences were still entirely inadequate, and the main burden of night fighting still rested at this stage with the day squadrons.

Operating on a wavelength of 1.5 metres, in the VHF frequency band between 190 and 195 MHz, the transmitter aerials were fixed in the nose of the aircraft, which worked on a floodlighting, or non-directional, principle. The arrow-head had two barbs, the front being the transmitting aerial, the second was a reflector to concentrate maximum power to the front. The receiving dipoles above and below either wing measured the difference in return time of any echoes back from a large target, such as another aircraft, or the ground. Four receiver aerials were arranged in two pairs, a pair to pick up the returned echo above and below the wings, with the other pair on the wingtips.

The AI Mk IV scope was suspended from the roof in the rear of the

4 B Robertson, "Airborne Interception", in *Aviation News*, 31 Dec 1993-27 Jan 1994, p.34.

5 Richards, op cit, p.203.

6 Gunston, op cit, p.54.

By November 1940, there were five squadrons of AI Mark IV equipped Beaufighters operational.

The 75cm arrow-head transmitter array on the nose of a Beaufighter If, and the vertical receiver pair, on the upper and lower wing surfaces, which derived target elevation. The wingtip pair received in the horizontal plane to resolve target azimuth.

The RO's two cathode ray tube display for the AI Mark IV, with visor, installed in a Mosquito.

Beaufighter, just behind the perspex dome, for operation by the radar operator (RO). The display could see behind the aircraft as well as in front, and it was not always immediately apparent to the RO from the cathode ray tube whether the target blip was in front or behind the fighter. However, the main disadvantage of the long-wave AI Mk IV was that it could not track a target at low altitude because of interfering ground reflections. This, in effect, restricted radar range to the height of aircraft above ground.

Target responses, blips, were then displayed on two cathode ray tubes, one showing angle off in azimuth with range to target, and the other showing relative elevation and range. Of course, because of the floodlight transmission, the large ground return swamped any small aircraft return at a greater range than the aircraft's height. If the aircraft was 15,000 feet above the ground, the maximum radar range was, therefore, 15,000 feet (3 miles).

AI radar was first revealed to the Americans in September 1940, and it was one British radar application that was totally new to US electronic research. AI Mk IV became the first airborne radar installed in an American aircraft. The US copied this design as the SCR-540, but through the latter half of 1941 and into 1942 had difficulty adapting it to their aircraft.[7] It was not until late spring 1942 that the SCR-540 began to enter service.

Meanwhile during 1942, although there was limited use of AI on Malta,

7 G R Thompson, *United States Army in World War II, The Technical Services - The Signal Corps*, Dept of the Army, Washington, 1957, p.86.

generally the Beaufighters sent to the Mediterranean were stripped of AI, to enable the home defence aircraft to be equipped. One such unit ill-equipped was No 255 Squadron RAF. The diarist of the Operations Record Book obviously felt the loss of AI, when he recorded:

> There is nothing more galling than to fly about near the flak, with parachute flares dropping from the Hun aircraft, on a bright moonlit night, and yet see absolutely nothing owing to being without radar. The Huns would have been sitting birds on these nights if only our AI equipment had been installed in the aircraft.[8]

However, from December 1942, AI equipment did start arriving from UK and became standard during 1943, when GCI control was also set up. Similarly, AI Mk IV became standard for the RAF in the Far East over 1943-44.

AI Mark V

FIU experimented with a pilot's indicator tube from October until the end of 1940. This feature was a distinguishing feature of AI Mk V, which was to become the standard radar for the next generation of night fighter, the Mosquito NF.II. AI Mk V was virtually a Mk IV, with the additional scope for the pilot showing a composite display of the navigator's scopes, on an indicator on the left side of his instrument panel. The Mark V utilised the same aerials as the Mk IV. The extra display for the pilot was so that he could carry out the final stages of the interception himself. The radar operator would scan for the initial contact, and the pilot could take over once the chase was on.

The first batch of Mosquitos were not delivered with AI Mk V until December 1941, when No 157 Squadron formed as the first Mosquito night fighter unit. By March 1942, the new squadron had received 14 Mosquito Mark IIs, equipped with the AI Mk V. The following month, No 151 Squadron traded its Defiants for the new Mosquito, and gained the first Mosquito night success on 29 May when a German bomber was damaged over the North Sea.[9]

The Mk V was considered inferior to the Mk IV, and those coming from AI-equipped Beaufighters to the Mk V-equipped Mosquito found operating the Mk V was a retrograde step, "like going back to the divining rod!"[10] Mark V AI was plagued by serviceability problems, and even though a large number of sets had been made, eventually all were replaced. AI Mk V was withdrawn from the few operational Beaufighters by February 1943, but the Mosquito had to soldier on with this much disliked equipment until the end of the year. Mk V AI was eventually scrapped and replaced on Mosquitos with Mk IV.

8 No 255 Sqn RAF 540, cited in Richards & Saunders, p.258.

9 Bowyer, op cit, p.140.

10 Robertson, op cit, p.35.

The AI Mark IV/V arrow-head transmitter aerials on a Mosquito II.

AI Mark VI

AI Mk VI was an adaptation of the standard AI Mk IV, operating between 188 and 198 MHz. Developed during 1941 for the Defiant II and single-seater fighters, AI Mk VI entered service that August, but was not to see widespread use. In all, over 6000 AI Mk IV/V/VI sets were produced during the Second World War.[11]

AI Radar - The Tool of the Trade

Although AI Mk IV had been a great technological advance and enabled targets to be engaged at night, radar serviceability was to remain a problem for the entire war. Nevertheless, by early 1941 AI Mk IV was achieving notable success. During the full moon period of 12-14 March, three AI Mk IV squadrons destroyed 15 enemy aircraft, and probably destroyed a further four.[12]

Initially, AI Mk IV fighters were not allowed to cross the French coast, so as not to risk falling into enemy hands. As the more advanced "centimetric" Mk VII/VIII AI sets became operational, the Mk IV-equipped aircraft were allowed to be used over enemy territory. But that was not until 1944, when three years of advances had made this older generation equipment obsolete, and night fighters were equipping with the next technological leap in radar innovation.

11 Martin, op cit, p.7.

12 Fighter Command Form 540, March 1941, excerpt in RAAF Historical Section, Canberra.

Mastering the changes in this rapid development and applying the techniques of AI was the job of the radar operators, the ROs:

> These men understood the mysteries of the cathode-ray tubes, made the radar work, learned to read the range from the green line on one tube, to gauge the difference in height from the other tube, and to give their pilots practical instructions. They learned how to ignore the clutter from ground returns, to make the best of a bad picture, to distinguish between approaching their quarry head on and from behind, to tell their pilots how to make an easy, converging turn so that they finished behind the bomber and not well out on a flank.[13]

13 AVM Johnson, op cit, p.161.

CHAPTER FIVE

BEAUFIGHTERS

The development of a useful airborne intercept radar, in the form of the AI Mark IV, had conveniently coincided with the arrival of the Bristol Beaufighter, which was to prove an ideal night fighting platform. The AI provided the night fighter's main asset of surprise, allowing it to find the target and stalk its prey to get into a good firing position.

The sturdy Beaufighter was considered to be a great aeroplane to fly. It was fast, with a maximum speed of 324mph and it could climb to 15,000 feet in just over nine minutes.[1] Its endurance enabled it to cope with long standing patrols, and the four 20mm Hispano cannon, beneath the nose, with six wing-mounted Browning 0.303 inch machine guns, gave it a formidable punch. Furthermore, the Beaufighter was comfortable, with a splendid field of view from the cockpit. There was plenty of room inside the fuselage, and it had improved heating for the crew on freezing nights. The crew had easy access through two hatches in the bottom, one behind the pilot's seat, the other ahead of the radar operator's (RO) position in the rear:

> Ducking down, the pilot climbed up the ladder fixed to the hatch, and into the seat (back folded down); by the time lapstraps were done up, a following airman had locked the seat-back up into place and fed the shoulder straps over and the pilot was ready to start up.
>
> In the rear seat, with fixed Perspex cupola above, the RO faced his radar. On either side of the walkway forward to the driving seat were the four 20mm cannon, with ammunition boxes and feeds. He suffered noise

1 Max speed of 324mph was at 11,750 feet. The Beaufighter II had a service ceiling of 27,000 feet.

and fumes when they were fired. The six 0.303in Brownings in the wings (two port, four starboard) made the Beaufighter one of the most heavily armed fighters of the war.[2]

No 456 Squadron received its first Beaufighter, R2474, on 26 September, and by the end of the month had taken delivery of a further five. There was soon found to be a major disadvantage to the new variant of the Beaufighter, the Mark IIf, that 456 had received. The Beaufighter I had originally been powered by a pair of Bristol Hercules radial engines, but the Beaufighter II was fitted with two Merlins, as a holdup in the production of Hercules had led to the temporary adoption of the Merlin.

First flying in October 1940, the Merlins produced 200hp less per engine than the Hercules. Moreover, the Beaufighter II proved to have a greater problem with directional stability, which had already been experienced in the Mark I Beaufighter. The original Hercules engines had rotated the propellers anti-clockwise, when viewed from behind. The Merlins rotated clockwise. With the fin being offset slightly to counteract the torque of the Hercules, the Merlin-powered aircraft consequently became longitudinal unstable, causing the aircraft to swing viciously to the left on take-off. If this was allowed to develop, it was almost impossible to correct, and was to lead to many accidents on Beaufighter II squadrons. Several modifications were trialled, but eventually this stability problem was not solved until the introduction of twelve degrees of dihedral to the tailplane. The take-off technique developed was to open the left throttle first, lifting the tailwheel off at 50mph to obtain directional control from the rudder, and continue increasing power with the left throttle about 2" ahead of the right. With the left throttle against the stops, the right throttle was continued forward, and the aircraft would fly itself off.[3]

Training on the Beaufighter was intensive over October and November 1941, despite poor weather, and there were to be a number of accidents as the pilots converted to the new type. On 15 November, Sergeant Jack Kirkman (T3024) misjudged his landing approach, and struck the sand dunes. He landed on one wheel and a wingtip, with the aircraft sliding along on the bottom of its engine nacelles, breaking the propellers and wiping off the undercarriage. Three days later more trouble occurred, when he bounced an aircraft (R2334) on landing, and on opening up the throttles, only had the starboard engine respond. The aircraft cartwheeled to the left, with both engines being torn out of the wings and the nose crushed in. The crew were lucky to be alive, escaping with only minor abrasions and bruises. That was enough for Jack Kirkman - he was posted to single-engined fighters, and saw out the war flying Spitfires.

A third accident occurred two days later, on 20 November. Pilot Officer

2 Wg Cdr J Meadows DFC AFC AE, "Flying Was Still Fun", Part 2, in *Aeroplane Monthly*, October 1992, pp.46-7.

3 Fred Stevens, interview with author, 11 Dec 1994.

Sergeants Jim Telford and Fred Stevens relaxing beside their Beaufighter II, T3028, with the early 456 Squadron code SA-V, at the beginning of 1942

Tom Cleary, accompanied by Sergeant Tom Condon[2], was landing his Beaufighter (R2372), when it swung to the left, and was pointing straight at the wreckage of T3024. The pilot opened the port throttle to correct the swing, and as the aircraft straightened up, the port undercarriage oleo leg collapsed. The aircraft slid along the runway writing-off the undercarriage, wing tip and propeller. As the Squadron diary recorded: "Third crash in Squadron in five days. Score now 3 for 0."[5]

It had been recognised that pilots needed to be adequately converted to twin-engined aircraft before arriving at 456 Squadron, and although this was now being done to some extent at the training units, it was still the role of the squadrons to convert crews to type. 456 Squadron pilots who had been flying Defiants, including Sergeants Fred Stevens[6] and Cliff Satchwell[7], were attached to 54 OTU, at RAF Church Fenton, Yorkshire, to complete twin-engine conversions on Blenheims. Even pilots arriving well into 1942,

4 Flt Lt T Condon DFC, 404399. No 1454 Flt RAF, 456 Sqn, 51 OTU, 264 Sqn RAF, 456 Sqn, HQ Ftr Cd. Grazier of Dalby, Qld; b Dalby 30 Aug 1914.

5 No 456 Sqn Operational Record Book, RAF Form 540, 20 Nov 1941.

6 Flt Lt F S Stevens DFC, 400739. No 456 Sqn, 54 OTU, 456 Sqn, 54 OTU. Leather worker of Melbourne; b Melbourne 7 Feb 1918.

7 Flt Lt C A Satchwell, 400736. No 456 Sqn, 60 OTU, 5 OTU, 30 Sqn. Clerk of Melbourne; b Melbourne 3 Feb 1920. Killed in action 13 Jul 1944.

One of 456's Blenheims, L4907 (coded '20' from its previous service at 54 OTU), used for twin-engined training.

An early shot of 456's Beaufighter at Valley.

such as Sergeant Gordon "Gate" Gatenby[8], who had converted to Blenheims at 51 OTU at Cranfield, had not flown Beaufighters before arriving at Valley.

8 Flt Lt G F Gatenby, 403330. No 456 Sqn, 51 OTU, 1 OTU. Pleasure ground proprietor of Bateman's Bay, NSW; b Sydney 20 Jun 1919.

456 Squadron aircrew off duty at the Menai Straits, on the Isle of Anglesey. From left: Tom Condon, Ron Lowther, Red Ratcliffe and Cliff Satchwell.

One improvement introduced on 456 Squadron's Beaufighters during November was with the arrival of two new aircraft (T3042 and T3047), which were the first to be delivered to the Squadron with belt fed cannon. The four cannon had initially been fed by drum magazines, which could be changed while airborne by the RO. Each drum held 60 rounds, and was therefore heavy and awkward to change, especially under the gravity forces of manoeuvring flight. During 1940, belt-fed systems had been trialled, enabling the introduction into RAF service from September 1941.[9]

The first Beaufighter operational sortie was flown on 29 November, and this, like a similar scramble the following evening, did not make contact, as the enemy raids did not venture as far north as the Valley sector. During December, 17 operational sorties were flown, mainly under the control of the local Valley sector GCI station at Trewan Sands, which used the callsign "Soapy". The 456 Squadron callsign was "Peeper."[10] Sometimes attempted interceptions took the Beaufighters to the adjoining Atcham Sector to the east, under control of Hack Green GCI (callsign "Gropa"), or to the north to Woodvale Sector to work under St Annes GCI control (callsign "Nofide"). Even if diverted to work outside the Valley sector,

9 Moyes, op cit, p.6.

10 "Peeper" was used by 456 over 1941-42, while operating from Valley. Known callsigns were: Peeper 15 - Merrifield, 17 - Timbury, 20 - Griffith, 22 - Newell, 26 - Gatenby, 31 - Pargeter, 36 -Stevens, 39 - Ratcliffe, 40 - Satchwell, 41 - Panitz. In January 1943, the 456 Sqn callsign changed to "Snooker" (Howard was Snooker 19), and changed again when 456 left Valley, to "Passport" (Bridges was Passport 27, Oxlade 33).

Working on the Merlin of a Beaufighter II. Ron Vidler front left, Ted Platt rear right.

Beaufighter II at Valley, March 1942. From left: LAC Abbot, Sgt Stevens (pilot), LACs Molloy and Gibson, Cpls Cooper and Thomas.

aircraft would always attempt to return to home base at "Bedside", callsign of Valley aerodrome control.

These first GCI sites were quite basic installations:

> The early mobile sites were simply cathode ray tubes and a few electrical gadgets mounted in a caravan with the rotating aerial on top. This aerial had a wooden frame, fifteen feet square and capable of rotation through 360 degrees. The wiring was such that after one complete revolution, it had to be reversed the other way. Power was supplied by an airman who sat on a bicycle seat and pedalled furiously. A warning bell rang when he had to back pedal. The remarkable thing about it all was that it worked.[11]

Clarity of radio communication, which was to be so vital for night fighters, had only recently been improved. Very-high frequency (VHF) radio telephony (R/T) was a new innovation that was being fitted to most aircraft in 1941, and this provided clearer communication between the air fighters and their GCI partners. Previously, R/T had been on unreliable HF frequencies, normally being faint and indistinct with a constant background crackle. The new VHF sets had four channels, but most aircraft were soon fitted with two sets, allowing for selection of eight channels, made by a push button control.[12] One RAF crewman from night-fighting Beaufighters remembered:

> It had more advanced VHF R/T receivers and transmitters which made communications with the GCI controller and aerodrome flying control much easier, with clarity of speech and less interference.
>
> Above all, the Beaus were fitted with the newly developed Mark IV AI set. Designed by our incomparable scientists, this miraculous equipment was compressed into half a dozen "black boxes", including a viewing unit into which the operator peered at two small cathode ray tubes. On the nose and wings of the aircraft was a neat aerial array to transmit the radar pulses and receive back the vital "echoes" as the pulses bounced back off any other aircraft within the coverage of the set.
>
> The Mark IV set flown in a Beau at 18,000 feet could pick up an "echo" in front of it three miles away. Its forward range was governed by the height of the Beau above the ground, ie at 20,000 feet the range was four miles but at 10,000 feet only two miles. Apart from the range, the signals (or "blips") received on the cathode ray tubes gave indications of the azimuth and elevation positions of the target aircraft relative to that of the pursuing fighter.
>
> A skilful GCI controller could, by watching carefully the blips of the incoming bomber and the defending fighter merging towards each other

11 Stevens, op cit, p.37.

12 The first VHF set allowed selection of the four main frequencies: A aerodrome control, B sector control, C emergency frequency (callsign "Darky"), and D local GCI. The second box, fitted to 456 Sqn Beaufighters from December 1941, allowed for selecting the frequencies of the adjoining sectors.

Gordon Gatenby with groundcrew.

456 Squadron Beaufighter II in its dispersal at Valley, about April 1942, displaying the Squadron's new 'RX' identification marking.

> on his single large tube (PPI), direct the fighter to within 2-3 miles of its target - sometimes less. Ideally he would attempt to put the fighter in on a "cut off vector" at the same height, or a bit higher, as the target, where the pursuer could turn in behind but not have a prolonged stern chase. Once the echo from the target appeared on the airborne radar, the AI operator shouted "Contact!" gleefully to his pilot, and took over, whilst the pilot passed on the good news to the GCI. From then on it was up to the crew to bring the interception to a satisfactory conclusion.
>
> Either by direct instruction or by a running commentary (or a combination of both), the AI operator could, after a swift but vital scrutiny of the signals on his tubes, aim to bring his pilot to within 5-600 feet astern of the target and slightly below. The shimmering green lines of the time tracks and the sometimes indistinct shape of the blip were not always easy to decipher whilst bumping along in the back of the Beau. Additionally, the amount of "working space" on the tubes, unencumbered by the menacing "ground returns", sometimes seemed woefully small (ie approx two centimetres at 10,000 ft). If the two aircraft were converging fast or even head-on, there were only seconds in which to act before the precious blip disappeared off the screen as quick as it had arrived.
>
> If all went well and the operator brought his pilot into close range he should start to see a shape which was different from the rest of night sky. On all but the darkest nights he would be able to pick up the outline of the other aircraft and coming in below, he would see the silhouette as it became larger and spread out above him but still slightly in front. On the way in he might have caught a glimpse of twin exhaust flames and if on final approach he chanced to pull up on the same level as his prey, he would feel the turbulence of its slipstream.[13]

A standardised patter was also developed by the night flyers for interceptions. In some crews, the navigator provided a running commentary of the target, explaining his AI indications, and the pilot interpreted the description to carry out the interception. On the other hand, some crews employed a navigator controlled intercept, whereby the navigator directed the pilot by instructions, as well as describing a clear picture of what was occurring. The ultimate aim was the same: to get into a position for the pilot to visually acquire the target against the lighter background of the sky. From the minimum range of the AI, this required the pilot to strain for that first flicker from the flames of the exhaust, and then close on the silhouette for a positive identification. Closing from below was done purposely, as it presented any rear gunner in the enemy aircraft, or even a trigger-happy gunner in a friendly bomber, a difficult search against the black background of the ground. A cloud cover below a moonlit night was particularly dangerous for the night fighter, but it was always necessary for visual recognition before attack. Positive identification of the target was paramount.

13 Mike Allen, in Franks, op cit, p.177.

Sergeants Mark Stokes (RO) and Lewis Parkes (pilot)

LAC McTear (fitter IIE), May 1942

Pilot Sergeant Gordon 'Gate' Gatenby, astride the Merlin of the CO's Beaufighter RX-A, T3383

5 January 1942 was not a good evening at Valley, as 456 Squadron was to have three accidents. The first was a tragic crash of Beaufighter (R2474), which resulted in the loss of two lives. Flight Lieutenant Tony Parker[14], a newly-arrived flying instructor, who had seen pre-war service in the permanent RAAF, and an American serving with the Squadron, Second Lieutenant F W Fraser[15], were carrying out an air test, with practice AI homings to Valley. The aircraft stalled, and spun into the sea on one of these approaches. The subsequent Court of Inquiry determined the cause to be an error of judgement in endeavouring to execute a port turn with insufficient flying speed, thereby stalling the aircraft at a height too low for recovery.[16] The wreckage of R2472 subsequently would sit in the shallow water as a reminder to pilots not to let the speed get too low on approach.

Three hours later, almost at 2000 hours on a dark night, Pilot Officer Tom Cleary (T3028) landed short of the runway threshold and struck the Glide Path Indicator (GPI) lights with his tailwheel.[17] The GPI was designed to provide a colour approach lighting system, indicating to the pilot the correct glideslope of his approach. It showed amber if too high on the approach, green if on the right approach, and red if below the glidepath. The system was subsequently modified, placing the lights either side of the runway instead of on the centreline, to avoid damaging the indicators when aircraft landed short.

The third accident that evening was an hour later. Sergeant Ray Hedge[18], flying Beaufighter R2476, was configuring for an approach into Valley and selected his flap setting. One flap failed, and the aircraft banked violently to starboard. Hedge had insufficient time to correct, and the aircraft crashed near Trewan Sands, about three kilometres from the base.[19] The crew escaped, but R2476 was a write-off.

The tide finally turned for the Squadron on the night of 10 January 1942, with 456's first taste of victory. All of 9 Group's airfields, except Valley, had been closed by the weather. The 'A' Flight Commander, Squadron Leader Hamilton and Pilot Officer Dan Norris-Smith (T3014), were scrambled at 2210 hours, and vectored by Soapy GCI towards an enemy raid. The contact was lost when Soapy experienced a radio failure, and vectors had to then be relayed through base operations. Eventually the adjoining Gropa GCI took control of the intercept, and after a seven minutes chase the

14 Flt Lt A R Parker, 250. Nos 1, 21, 4 & 6 Sqns, 1 SFTS, 143 Sqn RAF, 456 Sqn. Regular air force offr of Melbourne; b London, Eng, 26 Aug 1914. Killed in flying accident 5 Jan 1942.

15 Fraser was one of three US Army Signal Corps who had arrived on 456 Sqn at the beginning of December to gain experience with AI radar. It is interesting to note that they were posted to 456 as signals officers on 2 December, almost a week before the US entered the war.

16 Aircraft accident card for R2474, RAF AM Form 1180, RAF Air Historical Branch, London.

17 Aircraft accident card for T3028, RAF AM Form 1180.

18 Sgt R F Hedge RAF, 518431. No 456 Sqn.

19 Aircraft accident card for R2476, RAF AM Form 1180.

Beaufighter caught its prey. The target was identified as a fast and manoeuvrable Dornier Do 217 bomber, and was dispatched in flames after two bursts. The raider, from the *Luftwaffe* bomber wing *Kampfgeschwader* 40, crashed at Arbury, in Merseyside. The endless night training and practice intercepts had finally paid off.

Another accident occurred on 28 January, when Sergeant Gwynne Hughes, flying T3047, undershot the runway at night. He had been high in the GPI amber light, so throttled back and eased the stick forward. Now with a considerable sink rate, the aircraft lost height and hit the ground, then bounced onto the runway. The undercarriage collapsed and the aircraft swung off the runway into a sand dune. No one was hurt, and the Beaufighter was repairable. However, it did not prevent the Squadron diary recording: "B Flight's score 8 for Nil."

Daytime interception was now to become an additional task for the Squadron. In addition to 456 Squadron defending the Valley sector, was a Spitfire unit, No 350 Squadron.[20] In daylight, when weather was unsuitable for single-seat fighters, Fighter Command now scrambled the AI-equipped night fighters. Relying on the crews' instrument flying ability, which was essential to survival at night, it was hoped that there might be some success with this all-weather role.

The efforts expended over the following months continued towards training as the sector remained quiet. In addition to practice firing on an airborne drogue target pulled by one of Valley's Lysanders, a new innovative form of training was with camera guns. This form of cine training was to prove most beneficial in improving pilots' air to air firing. By the end of February, 456 Squadron had 15 crews declared as night operational, and another eleven under training.

Tom Condon and Cliff Satchwell, with Tom's car at Valley.

20 No 350 Sqn was replaced in March by 131 Sqn, which in turn was replaced in April by 315 (Polish) Sqn. An air/sea rescue unit, No 275 Sqn, equipped with Walrus amphibians and Lysanders, also shared the base.

Any aircraft that did get scrambled was inevitably to return empty-handed from a raid that never materialised. Whenever weather permitted, sorties were flown to assist the ROs in becoming more proficient with the AI equipment. One problem with AI radar was squint, where the aerial alignment was not properly boresighted to the aircraft's longitudinal axis. This meant that a target that appeared to the RO as being below his aircraft, could very well be above. In such cases, intercepts were never going to be made. It therefore became standard practice on night fighter squadrons that before on the night program, an aircraft would always be flown in the afternoon on the night flying test (NFT), to check aircraft serviceability and ensure the RO was aware of any squint in the particular AI.

An additional task for the RO was that of airborne armourer. He sat under the perspex dome in a swivel seat, facing rearwards towards the green glow of his AI scopes. He could rotate if he wanted to see forward, and with the drum fed guns, he was required to move forward and reload the cannon. As these were replaced by the belt-fed system, the RO was on hand if the guns needed attention.

Constant practice for the crews was necessary for them to become familiar with operating in the night environment and to develop their techniques. An improvement to assist in getting the aircraft back on to the ground had been the introduction of the standard Drem airfield lighting, named after the Scottish aerodrome where it had been developed:

> It consisted of an outer perimeter of low powered lights a half mile apart. At the threshold of the duty runway, a "spur" led to the start of the runway. This was known as the "funnel". Today it is short final. In poor weather it was terrific, simply follow the perimeter round, then down the funnel. The runway lights were hooded, and could not be seen above 3,000 feet but were quite bright on approach.[21]

Accidents had continued, including a spectacular incident on 20 March. Sergeant David Spring[22] was taking-off on his third solo flight in a Beaufighter (T3012), having completed some twin-engined training with 54 OTU at Church Fenton. He had been given some wrong advice for Beaufighter take-off technique, having run up to full power while holding the aircraft with brakes. On releasing the brakes, the aircraft began to swing off the runway towards the sand, so he pulled back a throttle to straighten and stay on the runway, only to have the aircraft swing savagely in the opposite direction. He then closed the other throttle, but by that time was already screaming down the runway, and so he attempted to reef the aircraft off the ground. The next thing into view was the windsock, which was struck by the port wing. The windsock post, a very sturdy piece of steel tube, ripped off the Beaufighter's port wing, causing the aircraft to skid

21 Stevens, op cit, p.29.

22 Flg Off D A H Spring, 402816. No 456 Sqn, 67 & 5 Sqns RAF, 3 TAF Comm Sqn. Salesman of Sydney; b Sydney 10 Aug 1914.

The wreckage of David Spring's Beaufighter II, T3012 RX-K.

sideways along the ground for about 300 metres, before bursting into flames. Initially uninjured, Spring tried to escape through the top of the canopy, but this hatch was jammed. With the front half of the aircraft on fire, he leapt through the flames that had destroyed the side of the cockpit, and ran alight out of the wreckage. He fell in the gorse beside the runway, to have someone fall on top of him with a blanket to douse the flames.[23]

Suffering from burns to the face and hands, Spring was to spend a month in hospital at Cosford. The aircraft was a complete write-off, providing a fireworks display as cannon shells and tanks burst. By the time he was ready to recommence flying, Beaufighter training was a little more formalised, and he completed a Beaufighter conversion with 54 OTU, which had now moved to RAF Charter Hall, Berwickshire in England's north.

From April, both flights of the Squadron commenced ground attack practice. Headquarters No 9 Group had directed that in the event of an enemy invasion, night fighter squadrons would be called on to strafe ground targets. A change in command was also to occur at the beginning of the month. Wing Commander Olive, who had been transferred through illness to Cosford Hospital at the beginning of March, had handed over to the 'A' Flight Commander, Squadron Leader Hamilton. Illness was to prevent Olive returning to operations, and he eventually was to depart for Australia. In his place, another RAF officer, Wing Commander Ted Wolfe[24], took over 456 Squadron on 2 April.

23 David Spring, interview with author 30 Dec 1994.

24 Wg Cdr E C Wolfe DFC RAF, 37705. Nos 64 & 219 Sqns RAF; comd 141 Sqn RAF 1940-41, 456 Sqn 1942-43; 9 Gp; comd 132 Wg RAF 1945. Regular air force offr of London, Eng; b Hong Kong 11 Jun 1911.

One of 'A' Flight's Beaufighter IIs, T3017 RX-B, which served almost a year on 456 Squadron before replacement by the Beaufighter VI.

Sergeant Gordon Gatenby assisting armourers to load the .303 machine-guns in Beaufighter RX-L, R2344.

It was at the stage that the Germans stepped up the tempo of their night attacks, as the *Luftwaffe* began its "Baedeker" raids against the county cities. This was in retaliation to Bomber Command attacks on Germany, and Hitler decreed that all British cities of architectural beauty, such as Norwich, Bath and York, would be removed from the Baedeker tourist guide! During the nights of 23 and 24 April 1942, about 50 German aircraft attacked Exeter each night, and the following night some 75 raiders attacked Bath.

Raids of this scale were not to venture into the Valley sector, and from early May the Squadron commenced daylight convoy escort patrols over the Irish Sea. It was on such a sortie that the second success came to 456 on Monday 18 May, for one of the RAF pilots, Pilot Officer Bernie Wills[25]. Flying T3022 with his RAAF RO, Sergeant Ron Lowther, who had been one of the Squadron's early air gunners on Defiants, he was tasked with a convoy patrol, to the east of Bardsey, the Welsh island off the western tip of Gwynedd. It was just after 1600 hours, as they were patrolling at 2,000 feet, when they spotted a Junkers Ju 88 amongst the cloud. For twenty minutes they played "cat and mouse" in and out of cloud, until they were eventually able to close. Despite the bomber's return fire from its upper dorsal turret, Wills was able to fire three accurate bursts, and saw his shots strike home. The Ju 88 was again lost in cloud, and so the crew had to be satisfied with a "damaged" claim. However, the Ju 88, from the *Luftwaffe's* long-range reconnaissance *Gruppe* 123, was later confirmed to have crashed in Cardigan Bay, and the crew picked up in their dinghy. It was now "One by Day, One by Night."

Problems in training with the temperamental Beaufighter II had continued. The accident with Beaufighter R2381, on night of 2 May 1942, was typical, being due to the pilot's lack of experience. The normal take-off swing had been uncorrected, and the aircraft had run into one of the sand dunes on the edge off the runway. The Station Commander was to comment: "To continue training the inexperienced pilots we are receiving would be a waste of life, aircraft and national effort."[26]

The good news was received during June that the unit would re-equip with the Hercules-powered Beaufighter VI in the near future. The Squadron diary recorded: "It is rumoured that having the radial engines for which the type was originally designed, these aircraft are slightly better to fly - having less tendency to swing when taking off and landing."[27]

On 26 June, Pilot Officer Phil Dey[28] and Sergeant E Mitchell[29] (T3014) were scrambled at 1145 hours to intercept a raid, 50 kilometres to the west

25 Wg Cdr D B Wills DFC RAF, 115997. No 456 Sqn, 68 Sqn RAF; comd 46 Sqn RAF 1960-61, 23 Sqn 1961-62. Export clerk of Heswall, Cheshire, Eng; b Woodchurch, Cheshire, 16 Apr 1922.

26 Aircraft Accident Card for R2381, RAF AM Form 1180.

27 No 456 Sqn 540, 15 Jun 1942.

28 Sqn Ldr P A Dey, 402853. No 456 Sqn, 276 Sqn RAF, 55 OTU, 100 Sqn. Audit clerk of Sydney; b Sydney 13 Jun 1921. Killed in action 16 Mar 1945.

29 Sgt E W Mitchell RAF. No 456 Sqn.

of Bardsey. The raid turned out to be a Junkers Ju 88, which in the ensuing combat put a hole in the balance pipe between the Beaufighter's two inner fuel tanks. The cockpit began to fill with petrol, and as the port engine stopped, Dey tried to make base on one engine. Unable to coax his Beau that far, Dey was forced to ditch. Making it clear of the sinking Beaufighter into a dinghy, they were rescued by a Walrus amphibian of the air/sea rescue squadron from Valley. Phil Dey was later killed while flying Beauforts with 100 Squadron in New Guinea.

Wing Commander Ted Wolfe DFC

The wing-mounted receiver aerials of the Beaufighter's AI Mk IV, which measured target elevation.

CHAPTER SIX

THE MARK SIX

Late in 1941, with the availability of the 1600hp Hercules 6 radial engine, the Beaufighter Mark VI was placed into production. Deliveries commenced to RAF squadrons in early 1942, and during July 456 Squadron received its first Mark VI aircraft (X8251). By the end of the month, fourteen of this new variant had been taken on charge. The additional power from the Hercules engine produced a quicker take-off, and when later aircraft were modified with a dihedral tailplane, the earlier deficiencies and problems of the Beaufighter Mark II were overcome. The pilots were soon pleased with the new aircraft.

The Beaufighter Mark VI was given its baptism to fire on the Squadron in quick time. On 29 July, five night patrols were flown by 'B' Flight, and while most aircraft did not pick up any activity, Wing Commander Ted Wolfe and Pilot Officer A Ashcroft[1] (X8251) were in the right area to obtain four contacts. Under Trewan Sands GCI control, the CO patrolled along the coast between Valley and Llanbedr, at medium level (up to 16,000 feet). The presence of many friendly targets, being plotted as "bogeys" (unidentified) made target selection difficult, and two of the intercepts proved friendly. However, another was a Ju 88, which took evasive action and was lost, while the fourth was a Heinkel He 111 of KG 40, which was shot down and crashed on the beach at Pwllheli, on the Lleyn Peninsula, 40 kilometres south of the base. The CO's Combat Report read:

> Closing in I obtained a visual at 2,000 feet range, identified the E/A [enemy aircraft] as a Heinkel 111, the exhausts on each side of the

1 Flt Lt A E Ashcroft RAF. No 141 Sqn RAF, 456 Sqn, 54 OTU.

Wreckage from the CO's Heinkel - Flg Off Tom Condon, Flg Off 'Peter' Panitz, FSgt Fred Stevens, FSgt Cliff Satchwell and Sgt 'Red' Ratcliffe.

> engine being very apparent. I closed in at IAS 270-280 on E/A now descending slightly to port and fired a 2-second burst with a slight port deflection at 250 yards followed by a second burst of 1-second at 150 yards. No return fire resulted, the upper gunner, having been shot through the head, the pilot's controls lost and the port engine put out of action during the first burst delivered. He 111 now skidded off to port jettisoning its bomb-load of incendiaries and I broke away to starboard closing in again to fire a 2-second burst from 50 yards with no deflection and breaking away violently to avoid collision. From this burst bright flashes, and falling fragments from the fuselage resulted. A large dark object passed beneath Beaufighter slightly to port. This is now presumed to be the port engine, as this engine has not yet been found in wreckage. The E/A was well on fire by this time. Both A/c were now losing height and I closed in to fire a final burst of 1-second at 150 yards range with slight port deflection, breaking off this last attack at a height of 4,000 feet, then climbed to starboard and levelled out. I saw He 111 burning in fuselage and port wing and continuing to lose height very rapidly. It turned slowly to port toward beach at Pwllheli...and when over beach at approx 2,000 feet dived vertically, exploding and burning on the beach.[2]

During July, Squadron Leader Geoff Merrifield[3] had taken over 'A' Flight

2 No 456 Combat Report of 29/30 Jul 1942, cited in No 456 Sqn Preliminary Narrative, RAAF Historical Section, Canberra.

3 Sqn Ldr G M Merrifield AFC RAF. No 456 Sqn; CFI 54 OTU 1943; comd 176 Sqn RAF 1945. Regular air force offr.

FSgt Fred Stevens' Beaufighter VI, 'Bunyip', the tyre cover on the wheel identifying the aircraft as X8138 RX-Z.

A close-up of Bunyip's nose art.

followed in August by a new 'B' Flight Commander, Squadron Leader "Bushy" Richards[4], who became renowned as "Handlebar Hank" for his flowing red moustache. This coincided with the arrival of the first large batch of radar-trained RAAF navigators at Valley, who up until this had been trained as observers, followed by a short course in the operation of AI. Now being trained as navigators (radio), to reflect the importance of airborne radar operation, this first group included Pilot Officers Dickie Williams[5] and Don Shanks[6].

The arrival of more RAAF aircrew was a deliberate policy to attempt to "Australianise" the Squadron as much as possible, allowing British crews to be posted to RAF squadrons. While RAF, and indeed later even Royal Navy, crews would normally be part of the fabric of 456, the Squadron was able from now to maintain a high proportion of RAAF personnel. It was also the intent of the EATS agreement, that when Australian personnel became available, they would command the RAAF squadrons, but in the case of 456 Squadron, an RAAF commander did not assume command until the end of 1943.

All 456's pilots were now solo on the Beaufighter, and the Squadron looked forward to getting all crews operational. While continuing their operational work-up, on 14 September Sergeant David Spring with Pilot Officer Jimmy "Tommy" Thomson[7] (X8205) experienced an engine failure. The starboard engine began running rough, and then seized, so the crew descended through cloud, attempting to return to Valley on the one good engine:

> As I cleared the overcast, I couldn't maintain height, so had to force land, wheels up in field. We hit one stone wall, then a second at the other side of the paddock, which brought the aircraft to a halt, the right way up, with everything hissing. A fire then started in the port engine, so I screamed out to Tommy to get out and run like hell. We both climbed out through our canopies, running about fifty yards, and jumping behind a wall as the Beaufighter blew up. The aircraft burnt, and by the time we were rescued, all that was left was a pile of white manganese dust in the profile of a Beaufighter.[8]

The crew were shaken up, but otherwise uninjured. A subsequent inquiry into the accident found that mysteriously there had been at least six

4 Sqn Ldr S P Richards RAF, 27112. No 29 Sqn RAF, 456 Sqn; comd 151 Sqn 1943. Regular air force offr.

5 Flt Lt R S Williams DFC, 411411. No 456 Sqn, HQ ADGB, 464 Sqn. Storekeeper of Patonga Beach, NSW; b Sydney 12 Aug 1915. Killed in action 22 Aug 1944.

6 Flt Lt D McK Shanks, 400839. No 456 Sqn, 10 Gp HQ, 464 Sqn. Bank clerk of Swan Hill, Vic; b Melbourne 6 Nov 1909.

7 Flt Lt J Thomson, 403374. No 456 Sqn, 151 Sqn RAF, 54 OTU, 5 OTU. Clerk of Sydney; b Sydney 1 Jun 1921.

8 David Spring, interview with author, 30 Dec 1994.

parachutes on board the wrecked aircraft. (Parachute silk was a luxury commodity to females in war-time Britain, so aircraft accidents became a convenient explanation for any misplaced parachutes.) It was naturally queried if it was indeed possible to get a Beaufighter airborne with so much extra weight on board!

The ROs' experience at operating AI had now shown marked improvement. During September, as the sector remained quiet with no enemy activity, GCI controlled interceptions were consistently leading to radar contacts, which were then being converted into "visuals". To ease the training burden on the Squadron, new crews arriving from OTU now were already Beaufighter qualified. When Squadron Leader Dot Dottridge had left 'B' Flight, he had taken up the post of Chief Flying Instructor (CFI) at 54 OTU at RAF Charter Hall, where most of 456's crews were now training. Well aware of the problems of converting crews to type on the squadrons, in addition to then bringing them up to an operational standard, he had ensured that adequate aircraft conversions took place at the OTU.[9]

Standard practice now too was for time-expired night fighter crews - those who had served on a squadron for eighteen months - to have their "rest" tour instructing on an OTU. This six months of "rest" was then followed by a further eighteen months on operations. For the RAAF night fighter crews, this invariably meant a second tour on 456 Squadron.

During the night of 8 October, Flight Sergeant Norm Scott[10] and Sergeant Grahame Wood[11] (X8190) were airborne for practice interceptions when their Beaufighter's starboard engine failed. Attempting to return to base, Scott requested the airfield's floodlights be turned on. The floods were designed to shine down on the runway threshold, when requested, to illuminate the first 200 metres of the landing strip. Scott then repeated his request, and that was his last transmission. The aircraft crashed at the "Q" site, a dummy airfield which was intended to deceive enemy raiders, situated about 15 kilometres south-east of Valley. The "Q" site had been lit up for an air defence exercise, and the crew had seen it through the clouds and mistaken it for their base. On the approach, the aircraft struck a sand dune, was completely broken up and burst into flames, costing them their lives.[12]

Although 456 had been operating the Beaufighter VI for three months, there was still a handful of the older Mark IIs retained for training. On 28

9 No 456 Sqn 540, 5 Oct 1942.

10 WOff R N B Scott, 400737. Nos 151 & 96 Sqns RAF, 456 Sqn. Shipping clerk of Melbourne; b Melbourne 4 Mar 1922. Killed in flying accident 8 Oct 1942.

11 Sgt G A C Wood, 401343. No 456 Sqn. Of Melbourne; b Hobart 10 Aug 1921. Killed in flying accident 8 Oct 1942.

12 No 456 Sqn 540, 8 & 9 Oct 1942.

Scotty Fraser and Gordon Gatenby in front of their Beau VI, 'Up 26!'

Sgt Lyle 'Skippy' Skippen

October, Pilot Officer Rex Timbury[13], with Sergeant Lyle "Skippy" Skippen[14], were landing a Beau II (T3383) at dusk, after an NFT. As the aircraft landed, it swung, and Timbury had trouble with the brakes trying to keep the aircraft straight. As the aircraft swung from side to side down the runway, another aircraft waiting to take-off on a cross runway quickly came into view. Timbury purposely swung his aircraft violently to the right towards a clear area, causing the undercarriage to collapse.[15] Neither crew were injured, and the aircraft was eventually repaired on 456 before being passed on to an OTU.

Three days later, there was an accident in more tragic circumstances as November opened badly. The "two Macs", Pilot Officer Doug "Ken" McGregor[16] and Sergeant Leo McCormack[17] (V8444), were carrying out practice interceptions with another Beaufighter over the Irish Sea on the night of 1 November. Just after 2100 hours, the other aircraft flown by Pilot Officer Geoff Oxlade[18] was changing the lead with Ken McGregor to continue the practice intercepts:

> Geoff had made the interception and advised Ken by radio that he was coming past on his right. As he went by, Geoff's navigator, Don Shanks, waved to the others. Then Geoff pulled ahead and Don watched as Ken began to turn in behind. As he watched, he saw the turn get steeper until the nose dropped and they disappeared. A few seconds later, a vivid explosion occurred on the water. Alarmed, Geoff advised Control. All we could think was he turned too tightly, stalled, and spun in.[19]

There was an electrical storm at the time,[20] and the accident was subsequently recorded as: "Aircraft missing. Aircraft flying at night. Aircraft probably struck by lightning whilst in or near cloud. 'An act of God'."[21]

Such effort was put into practice interceptions because of the complexity of operating AI, and also because it was necessary to execute the intercept geometry accurately. As GCI made handover to AI possible at 4 to 5 miles, the role of the AI operator was to get close enough behind the target for a

13 Flt Lt R S Timbury, 403835. No 456 Sqn, 68 Sqn RAF, 1 SFTS, 5 OTU, 87 Sqn, 2AD. Trainee chartered accountant of Tamworth, NSW; b Newcastle, NSW, 1 Jan 1917.

14 Flt Lt L T Skippen, 403115. No 456 Sqn, 5 OTU, 1 AOS, 7 OTU, 86 Wg HQ. Schoolteacher of Sydney; b Inverell, NSW, 28 Jan 1911.

15 Rex Timbury, interview with author, 2 Jan 1995.

16 Flg Off D L McGregor, 400725. Nos 151 & 256 Sqns RAF, 54 OTU, 456 Sqn. Engineer of Melbourne; b Euroa, Vic, 18 Feb 1917. Killed in flying accident 1 Nov 1942.

17 Sgt L M McCormack, 405183. No 456 Sqn. Schoolteacher of Gympie, Qld; b Gympie 4 Aug 1919. Killed in flying accident 1 Nov 1942.

18 Sqn Ldr A G Oxlade, 400733. No 256 Sqn RAF, 456 & 464 Sqns. Clerk of Melbourne; b Echuca, Vic, 8 Apr 1920. Killed in action 5 Jun 1944.

19 Stevens, op cit, p.42.

20 456 Sqn 540, 1 Nov 1942.

21 Aircraft accident card for V8444, RAF AM Form 1180.

visual acquisition. The turn called by the RO was crucial - if it was called to soon, the Beaufighter might roll out in front of the bogey. If the turn was too late, the crew was faced with a long stern chase. The best method of gaining an early visual spot was to approach from astern at a slightly lower level. In good visibility with bright moonlight, it might be possible to spot the target from 4000 feet, nearly a mile. More typically, a bomber could be seen from 1000 to 1500 feet on an average dark night. Without the advantage of starlight, it may even have been necessary to close to 600 feet until gaining a visual contact. The pilot would then be straining to make out the silhouette, and searching for the tell-tale flames of the engine exhausts, so that he could close to carry out his identification.

The importance of positive identification was always stressed, as it was considered paramount. Now radar technology was again to take a leap to aid the night crews in target identification, or part way, at least. This innovation was IFF - Identification Friend or Foe - which enlarged the radar echo on the scopes of the GCI or AI sets. The theory was that the intense blip, which would flash prominently, would identify the aircraft as friendly. The problem was that if an aircraft was not "squawking" IFF, it did not mean automatically that the target was hostile. It was a "bogey", being unidentified, which could be a friendly without the correct IFF setting, with an unserviceable set, or perhaps, had just forgotten to switch it on. Therefore, bogeys had to continue to be identified visually and confirmed as hostile, a "bandit", before it could be attacked.

Night fighter squadrons were already being called upon to make day interceptions in foul weather which prevented the single-seaters flying, and now 456 Squadron was tasked with early morning sector reconnaissance patrols. This involved looking for German U-boats and floatplanes off the southern coast of Ireland. The reconnaissance Beaufighter would plan to be at Carnsore Point, the south-eastern corner of Eire, at first light, to search the coast to Kinsale, south of Cork. This type of reconnaissance, which had been instigated from within 456 Squadron, became known as the "Irish Sweep". Taking off on one of these sorties at about 0630 hours on 3 November, Flight Sergeant Fred Stevens, with Flight Sergeant Mark Stokes[22] in X8138, recalled a banging as they sped down the runway, which they presumed was an engine backfiring:

> On return from the uneventful reconnaissance, as all engine indications had been normal, I decided to have a visual inspection by the control tower. One of our pilots, Phil Dey, was on duty and he radioed: "Your starboard tyre is rooted - it's flapping in the breeze!" My options were to land wheels up and belly land, or attempt a normal wheels down landing. I chose the latter, and we zig-zagged down the runway, to successfully emerge shaking! The doctor took me to base sick quarters, to be soothed with a shot of overproof rum.[23]

22 Flg Off M A Stokes, 402620. No 456 Sqn, 51 & 54 OTU. Motor mechanic of Sydney; b Sydney 24 Feb 1919. Killed in flying accident 23 Nov 1943.

23 Fred Stevens, interview with author, 11 Dec 1994.

Flight Sergeants Mark Stokes and Fred Stevens

Fred Stevens and Mark Stokes (X8194) flew another Irish Sweep on 19 November, in company with Flight Sergeant "Red" Ratcliffe[24] and Sergeant Alec Abbey[25] (X7895). After becoming separated, Stevens returned to Valley, and by late morning Ratcliffe had still not returned. Stevens and Stokes then departed with the 'B' Flight Commander, Bushy Richards, to conduct a search along the southern Irish coast. All they found was an oily patch in the sea off Carnsore, but then received by radio the heartening news that the missing crew were safe in Ireland. Ratcliffe and Abbey had lost their radio and radar, flown unintentionally over Eire, then forced landed near Omagh, in County Tyrone, Northern Ireland, due lack of fuel. The aircraft was slightly damaged, and the uninjured crew returned to Valley two days later.

No 456 Squadron was advised by 9 Group on 2 December 1942 of plans to re-equip with the Mosquito Mark II, and the following week crews began lectures on AI Mark V in anticipation of the new radar fitted to the Mosquitos. A fortnight later came the announcement that Wing Commander J McDougall DFC was to take over the Squadron from Ted Wolfe, but this was immediately prevented by his death in a flying accident. Within days his replacement was nominated as Wing Commander Vernon Wood[26], the CO of No 604 Squadron, who was an Australian serving in the

24 WOff V P Ratcliffe, 403373. No 456 Sqn. Clerk of Sydney; b Sydney 8 Oct 1919. Killed in action 2 Jun 1943.

25 Flt Lt A M Abbey, 403128. No 456 Sqn, HQ 10 Gp, HQ ADGB, 456 Sqn. Clerk of Coffs Harbour, NSW; b 26 Jun 1916.

26 Wg Cdr V C Wood AFC RAF, 37309. Comd 604 Sqn 1942-43. Regular air force offr of Terang, Vic; b 18 Jul 1913. Killed in action 17 Apr 1943.

The Beau VI, X8100 RX-E, showing its new colours, the green/grey camouflage replacing the overall black.

Cycling at Valley - 'B' Flight Commander, Sqn Ldr 'Handlebar Hank' Richards and Flg Off Bob Hitchcock

RAF. He had trained at Point Cook before the war and transferred to the RAF, but this appointment also was changed. Finally, Wing Commander Mickey Dwyer[27] joined 456 at Valley on New Years Eve as Commanding Officer-designate. His arrival coincided with that of the first three Mosquitos.

The Beaufighter had turned the tide of night fighting for the RAF. It marked the beginning of an airframe being matched with aircraft systems, to provide an effective weapons platform. From Defiants, No 456 had been armed with Beaufighters, and now in fairly quick succession, was rearming with Mosquitos. These changes had involved considerable and continuous training, which led to feelings by air and ground crews alike, that often the Squadron was no more than a "glorified OTU". Being based in Anglesey for the past eighteen months, with the marked disinclination of the Germans to visit the sector, had reduced any opportunities for operational activities. Now, armed with new aircraft, 1943 would hopefully see a change for the Australians from the backwater at Valley into more operational circumstances.

27 AVM M H Dwyer CB CBE, RAF. Comd 456 Sqn 1943, CGS 1943-45. Regular air force offr of Heswall, Cheshire, Eng; b London 18 Sep 1912.

CHAPTER SEVEN

MOSQUITO RANGERS

Designed and built in the ancient moated manor house in Salisbury Hall, Hertfordshire, to avoid possible bombing of de Havilland's nearby Hatfield factory, the Mosquito was conceived as a completely undefended all-wooden bomber. The project had been a private venture by the de Havilland company, and had to overcome official prejudice against the concept of a bomber with the speed of a fighter. The prototype Mosquito was moved to Hatfield, and first flew on Tuesday 25 November 1940.[1]

The initial production version, the PR.I for photographic reconnaissance, was quickly followed by a second variant, the F.II fighter. Equipped with AI radar, this became the NF.II night fighter, which flew in May 1941. The Mosquito was to become a true multi-role warplane, and in all, nearly fifty variants would be produced.

Entering RAF service in January 1942, the first Mosquito NF.II night fighters were accepted by No 157 Squadron at Castle Camps, a new base built near Cambridge as a satellite field of RAF Debden. They were soon joined by No 151 Squadron, at Colerne in Wiltshire, and standing patrols were mounted from June over the North Sea. These soon paid dividends – the first confirmed "kill" for a Mosquito NF.II was on the night of 24/25 June 1942, when Wing Commander "Black" Smith[2], the CO of 151 Squadron,

1 This prototype, serialled W4050, has been restored and is on display at the home of the Mosquito, at Salisbury Hall's Mosquito Aircraft Museum. It is dedicated as a memorial to all those who had designed, built and flown the Mosquito.

2 Gp Capt I S Smith CBE DFC* RAF, 43048. No 151 Sqn RAF; comd 151 Sqn 1942-43; HQ Ftr Cd, HQ 2 Gp; comd 487 Sqn RNZAF 1944, 13 OTU 1944-45; AHQ Malta; comd 56 Sqn 1950, Church Fenton 1952-53, Jever 1959-61. Regular air force offr; b Invercargill, NZ, 21 May 1917.

shot down two Dornier Do 217s and claimed a Heinkel He 111 as a probable. The Mosquito was to prove itself as a deadly hunter, equipped with Mark IV or V AI "arrow-head" radar and armed with four .303-inch Browning machine guns in the nose, which had been modified from the perspex nose of the PR.I. In addition, the NF.II also carried four 20mm Hispano cannon in the front fuselage belly. Just a 1½-second burst from the four cannon produced a deadly punch of 100 rounds, normally a mixture of HE (High Explosive) and SAPI (Semi Armour Piercing Incendiary) shells.

To make room for the barrels and flash guards of the four 20mm cannon fitted in lower fuselage, the entrance door of the PR.I (and later bomber variants) was moved from a hatch in the floor to the starboard fuselage side. Entry into the fighter was up a long rickety ladder, pilot first, as there was no room otherwise to squeeze through, which made scrambles a little slower than in Beaus. Also, after the spacious Beaufighter, the crew of the Mosquito were cramped with the pilot and radar operator side by side:

> Although the RO's seat was set slightly back (he had to lean forward to the visor on his AI display), elbows frequently clashed, particularly when the pilot operated the fuel cocks behind his armour-plated seat.[3]

The NF.II had a maximum speed of 370mph at low level, 407mph at 28,000 feet, with an initial climb rate of 3,000 feet per minute. With a ceiling of 36,000 feet, the aircraft's flight envelope was quite remarkable. Typical fuel consumption for economical cruising was 60 to 70 gallons per hour, which gave a maximum range of just over 1700 miles (2750 kilometres). One Mossie pilot remembered the aircraft's basic flying characteristics:

> Take-off was not nearly as difficult as I had been led to believe, and on the climb-out and approach the only things one needed to anticipate were a slight wiggle as the undercarriage came up and a yaw as the gear was lowered on the downwind leg.
> However, power on the final approach had to be used with care, and full flap was only to be selected when you were 100 per cent sure you were going to make the runway.[4]

Initially, RAF Mosquitos had been painted, like the Beaufighter IIs, in standard night fighter finish of Sooty Black[5], a rough finish which gave off no reflection. This was abandoned when found it cut some 26mph off the Mosquito's top speed.[6] The replacement Flat Black finish was shown to present a strong silhouette against clouds on moon-lit nights, so a semi-matt Medium Sea Grey/Dark Green scheme became standard for all night fighters.

3 Meadows, Pt 2, p.47.

4 Flt Lt M Retallack, "Mastering the Mossie", in *Aeroplane Monthly*, Oct 1993, p.40.

5 The official RAF colour designation was RDM/2A.

6 J Scutts, *Mosquito in Action Part 2*, Squadron/Signal, Carrollton, Texas, 1993, p.7.

DD737, an early Mosquito NF.II finished in overall Sooty Black

Cockpit of the Mosquito NF.II

The four deadly cannon mounted in the Mosquito's belly.

The introduction of the Mosquito by the Australians had also led to a very interesting marking for 456 Squadron aircraft. A flight rigger who had sailed with the original 455/456 Squadrons' Blue Draft had been Aircraftman Ron Vidler[7], who recalled:

> On our way over, we made templates of cardboard of the kangaroo from the penny. Most of the chaps around us on the ship emptied their blue kit bags and painted the "roo" on them. The Squadron went to the Isle of Anglesey...it was here you will see me standing with my back to the

7 Cpl R Vidler, 14800. Nos 455, 452 & 456 Sqns, 3064 SE, 6456 SE, 6451 SE. Mechanic of Lismore, NSW; b. Sydney 7 Feb 1920.

456's kangaroo marking on the crew entrance door of the Mosquito, the additional markings refer to kills of German aircraft and V1 flying bombs.

> camera and the roo on the overalls. We put the roo in the middle of the roundel when we got the Mossies. On my return to Australia, I wrote to Air Force Headquarters and said it would be a good idea to have it on RAAF aircraft – didn't get a reply, time went by and it appeared on RAAF aircraft.[8]

Indeed, in 1956, the RAAF did adopt the kangaroo roundel, and as the accompanying photograph of a Mosquito crew entrance door shows, today's RAAF roundel is exactly the same as that adopted by 456 Squadron in 1943.[9]

As 456 Squadron and other units equipped with the Mosquito over late 1942 and early 1943, the modifications suggested and techniques pioneered by the first two Mosquito night fighter squadrons were adopted. One improvement had been the fitting of flash suppressors to the machine guns to avoid blinding the pilot. An RO who made the transition from night-fighting Beaus to Mosquitos commented:

8 Ron Vidler, letter to author, 5 Dec 1994.

9 The RAAF only adopted the "kangaroo in motion", after a RAAF-wide competion in 1956 to select a national marking, on aircraft fuselages only. The British style target roundel was retained on aircraft wings. In 1965, the decision was taken to place the kangaroo roundel also on the wings of RAAF aircraft. Air Board Agendum No 13128, of 16 Sep 1965.

AC Ron Vidler with the leaping kangaroo on the back of his overalls.

> The main advantage for me was being able to sit at the side of the pilot rather than being stuck down the back and facing aft. One could see more of the action but I had to keep my eyes on the AI until the pilot was in the firing position. Sitting side by side also gave a better feeling of team work.[10]

By 19 January 1943, most of 456's pilots had soloed on the new aircraft. What was required now was for the navigators to become competent with the AI Mark V, which was assessed as requiring about ten hours of flying, before the crews could be declared operational. The first accident during this work-up period occurred on 25 January, when Flying Officer Roger Hitchcock[11] and Pilot Officer Les Parnell[12] (DZ658) landed about two miles short of Valley's westerly runway. The aircraft initially rolled on its wheels, but as these were broken off, it then skidded across the ground on its belly. After almost a mile, the aircraft went through a fence which ran beside the railway, with parts of the wreckage coming to rest on the track. The crew were uninjured, and clambered clear to walk back to base, with the Mosquito, of course, being a total write-off. About ten minutes later a train came through, and made its normal stop at Rhosneigr station, where the stationmaster remarked to the driver: "You have a tyre on your front buffer!"[13]

10 Terry Clark, in Franks, op cit, p.184.

11 Flg Off R B Hitchcock, 407714. No 456 Sqn. Service mechanic of Adelaide; b Sydney 2 May 1918. Killed in flying accident 23 May 1943.

12 Flg Off L R Parnell, 411178. No 456 Sqn. Of Sydney; b Sydney 5 Mar 1920. Killed in action 4 Dec 1943.

13 Fred Stevens, interview with author, 11 Dec 94.

The kangaroo then became a roundel, for display around the Squadron and, unofficially, on the aircraft.

Another accident occurred the following week, on 30 January. Sergeants Bill Anderson[14] and Les Gill[15], flying Beaufighter EL214, were acting as a target in a cine gun camera exercise, so that Warrant Officer "Gate" Gatenby could attempt some filming with one of the new Mosquitos. The target aircraft began weaving, so Gatenby requested to Anderson to maintain straight and level as briefed for the exercise. The Beaufighter then took even more violent evasive action. Pulling up from a height of 4000' at 260mph, the aircraft banked over the vertical into a wingover, rolled onto its back, and now with a nose down attitude, entered a steep dive. With vapour trails streaming from its wingtips, the aircraft accelerated down at a steep 45 degree angle, as the pilot fought the controls, attempting to pull out of the dive. The aircraft hit the sea in excess of 400mph, exploding on impact.[16] The aircraft accident board determined there had been insufficient height for recovery from the evasive manoeuvre, as the aircraft was flying 1000' below the minimum authorised for the exercise.[17]

On 1 February 1943, Wing Commander Mickey Dwyer assumed command of the Australian night fighter squadron, as operations remained at a low level, with few scrambles and little enemy activity. (In addition to being Commanding Officer of 456 Squadron, Dwyer had another duty, that

14 Sgt W R G Anderson, 405167. No 456 Sqn. Of Brisbane; b Brisbane 1 Mar 1917. Killed in flying accident 30 Jan 1943.

15 Sgt L Gill, 406279. No 456 Sqn, 54 OTU, 456 Sqn. Clerk of Perth; b Albany, WA, 15 Nov 1918. Killed in flying accident 30 Jan 1943

16 Gordon Gatenby, interview with author, 17 Dec 1994.

17 Aircraft accident card for EL214, RAF AM Form 1180..

Beaufighter VI EL214, the aircraft that crashed on 30 January 1943.

of CO of RAF Valley.) Opposition had been scarce, as the *Luftwaffe* was not daring to venture in large numbers now over England, and this did not look like changing. Towards the end of the month, on 25 February, one of the ROs, Flight Sergeant "Charlie" Stovell[18], died at the RAF Hospital at Cosford, from complications following a kidney operation. Stovell had been one of the Squadron's air gunners during 1941, and since becoming a navigator/radio, had often flown sorties with Wing Commander Wolfe.

However, there was now good news for 456's future operations – the Squadron was to be tasked with "Ranger" sorties. Night intruding had begun at the end 1940, when No 23 Squadron RAF had mounted night patrols over French airfields, penetrating deeply into enemy occupied territory on dark nights with little moon, and attacking German bombers at their own bases.[19] Intruder operations had become so successful, that by early 1943 it was decided to extend these to freelance operations against targets such as trains, lorries, barges and anything that moved at night. The aim of Ranger operations was to force the enemy to maintain and disperse a fighter force in the west at high readiness, destroy his aircraft, disorganise training schedules and attack transport.[20]

Night fighter squadrons were selected to conduct these Rangers, as night raids on Britain were on a very much reduced scale. Being conducted

18 WOff C Stovell, 404220. No 456 Sqn. Wharf labourer of Rockhampton, Qld; b London, Eng, 30 May 1908. Died of illness 25 Feb 1943.

19 J D R Rawlings, *Fighter Squadrons of the RAF and their Aircraft*, MacDonald Janes, London, 1978, p.59.

20 C M Sharp & M J F Bowyer, *Mosquito*, Faber, London, 1971, p.341.

initially only on bright moonlight nights in radio silence, the predetermined times of crossing out and returning for Ranger sorties were passed to the sectors so as not to be mistaken for enemy raiders. While the moonlight period lasted, a night fighter squadron interspersed Ranger operations in addition to long-range day fighters sorties over occupied France, with normal nightly patrols for home defence.

During March 1943, Nos 29, 157 and 456 Squadrons began their first Rangers. For these missions, a flight of an additional six aircraft were to be allotted to 456 Squadron, and at last, the Australians looked forward to some promise of activity. Training proceeded for the Ranger crews under the new CO during early March, which comprised flying long cross country exercises to develop skills in low-level navigation. Soon word was received from Headquarters 9 Group, that three aircraft and six crews would be required to be deployed to RAF Colerne, in Wiltshire, to be ready for Ranger operations from 16 March. The aircraft were prepared, by the removal of the AI radar, as this equipment was considered too classified to risk flying it over enemy territory.

Rangers started from forward bases, with 10 Group concentrating on western and southern France (which would include 456 Squadron), 11 Group the eastern part of France, Belgium and southern Germany and 12 Group Holland and northern Germany. The first Australian Rangers were flown on the night of 21 March from Colerne, by Pilot Officers Harry Smith[21] and Jack Ross[22] (HJ916), and Warrant Officer Red Ratcliffe and Flight Sergeant Ron Lowther (DZ307). Poor visibility in thick ground haze prevented the crews seeing any rail targets, and two nights later, three crews were forced back by weather before making landfall over France. On the 25th, orders were received that 456 was to move to Middle Wallop, near Andover in Hampshire, on 30 March, which would enable the Rangers to be operated from home base.

Middle Wallop was a permanent pre-war RAF station in 10 Group, and 456 Squadron was swapping places with a Canadian night fighter unit, No 406 Squadron. Squadron Leader "Robbie" Robinson[23], an Australian serving in the RAF, arrived before the move to take over 'A' Flight from Geoff Merrifield. Although there may had been a little regret to leave Valley, as the Squadron had grown up there, Middle Wallop held the promise of more activity.

Middle Wallop had originally been planned as a Bomber Command station, but was taken over by No 11 Group while still being built in 1940. The newly formed 10 Group occupied the station just in time for the first major assault of the Battle of Britain. The new airfield, although equipped with a concrete perimeter track, had a grass landing ground with wire mesh

21 Sqn Ldr H Smith, 403821. No 456 Sqn, 51 OTU, 456 Sqn. Carpenter of New Lambton, NSW; b New Lambton 24 Nov 1917.

22 Flt Lt J R Ross DFC, 401721. No 456 Sqn, 62 & 51 OTUs, 456 Sqn. Shipping clerk of Melbourne; b Warragul, Vic, 12 Nov 1915.

23 Gp Capt G M Robinson DFC RAF, 42151. Nos 219 & 23 Sqns RAF, 456 Sqn; comd 52 OTU 1944. Regular air force offr of Sunshine, Vic; b 8 Jul 1914.

'A' Flight Middle Wallop (left to right): Stewy Smith, Ivor Sanderson, Harry 'Smithy' Smith, Bob Richardson, Harry 'Hoppy' Williams, Gordon 'Gate' Gatenby (kneeling), Doug Wiley, John 'Scotty' Fraser, Stan Williams, Johnny Newell

laid to enable all-weather operation, which permitted a maximum run of 4,200 feet into the prevailing south-westerly wind. The hardstandings, not being dispersed like Valley, housed both flights in one hangar, with the servicing echelon and Squadron headquarters in the next. The officers' accommodation was off-base, some ten kilometres away in "The Pheasant" hotel, on the Andover Road near Salisbury. This was to be a very comfortable Officers' Mess.

Rangers recommenced for the next moon period on 16 April, when Wing Commander Dwyer and Flying Officer Don Shanks (DZ269) flew along railways in France, but saw no targets. Encountering intense flak west of Bayeux, they were able to avoid being hit, although two Beaufighters that had also departed Middle Wallop on Ranger sorties that night were both damaged by flak. The following night, Ratcliffe and Lowther (DZ269) had a more successful sortie, damaging a train which was brought to a standstill and left enveloped in steam.

Meanwhile, a state of six aircraft at readiness was required each night and normal night fighter interceptions were flown in the sector around Southampton and the Isle of Wight, under the control of Sopley GCI (callsign "Starlight"), on the south coast near Bournemouth. The crews were finding the reliability of the AI Mark V to be poor, and its reactions to manoeuvring targets proved to be slow. The known deficiency of the system of being hampered by ground returns was especially being exploited by the enemy, as the *Luftwaffe* was now making raids at low level. This was demonstrated on the night of 24 April, when Pilot Officer

'Peter' Panitz and Dickie Williams

Johnny Newell[24] and Flight Sergeant Allan Keating[25] (HJ938) were vectored after a Ju 88 in extremely bad weather, with a solid cloud base at 800 feet and steady drizzle. A fleeting contact at maximum range was unable to be held as they were only at 4,000 feet, and the crew returned empty-handed. This Hun was brought down eventually, however, by a 85 Squadron Mosquito fitted with the new mark of "gremlin box", the centimetric wavelength Mark VIII AI, which was virtually unaffected by ground clutter. Nearly a year would elapse before 456 Squadron would re-equip with a new AI, and hence the concentration now on Ranger work. As one of the navigators on 456 at the time, Merv Austin[26], recorded, the crews looked towards the Rangers with enthusiasm "...for the action denied them in night fighting".[27]

Night fighting was also planned at this stage to extend to several RAAF Mosquito squadrons in Australia, to counter the Japanese night bombing of

24 Flt Lt J W Newell, 404543. No 256 Sqn RAF, 456 Sqn, 63 OTU, 1 OTU, 5 OTU. Clerk of Mareeba, Qld; b Cairns, Qld, 8 Jun 1917. Killed in flying accident 5 Mar 1945.

25 Flt Lt A J Keating, 403654/O24189. No 456 Sqn, 51 OTU, 125 Sqn RAF, HQ EA, RSTT, 78 Wg HQ, 2 SD. Insurance agent of Sydney; b Coonamble, NSW, 8 May 1917.

26 Wg Cdr M N Austin, 409368/O55793. No 456 Sqn, 10 Gp HQ, O'Seas HQ; comd WA Uni Sqn 1960-68. Schoolmaster of Melbourne; b Melbourne 1 Aug 1913.

27 Flt Lt M N Austin, Preliminary Narrative of the History of No 456 Sqn RAAF, Air Ministry, London, 1945, Part III, p.5.

Darwin. One 456 crew, Flying Officers Rex Timbury and Lyle Skippen, teamed up with another RAAF crew from an RAF night fighter unit from the Mediterranean, and were attached to No 68 Squadron RAF at Coltishall, Norfolk, over April and May. The crews flew Mosquitos with the new Mark VIII AI, to gain experience with the latest technology in anticipation of Australian-produced Mosquitos receiving American radars. The crews then visited the USAAF base at Orlando in Florida, and flew with the American 348th Night Fighter Squadron in Havocs, before returning to RAAF Headquarters in Melbourne to prepare a night fighter syllabus. The night fighting OTU was proposed to be based in Wagga, but by September 1943, delays in the Australian Mosquito production program led to the urgency for the requirement to recede.[28] In the event, the RAAF was not to operate AI-equipped night fighters during the Pacific War.[29]

Day Ranger sorties commenced on 6 May, and 456 had the most successful Ranger flown by any Allied crew to date. Pilot Officer "Peter" Panitz[30] and Flying Officer Dickie Williams (DZ269) crossed the French coast at Mont St Michel at about 1500 hours that afternoon. Flying at "zero" feet, the conditions were ideal for a day Ranger, being protected from enemy air activity below an 800' cloud base. They successfully attacked a train leaving Combourg, then proceeded on a clockwise route around Brittany, centred on St Malo, shooting up a further five goods trains. The Squadron diary recorded: "A really grand trip." A postwar RAAF publication described the railway adventure in colourful terms:

> A single engine was on the line, leaving Combourg, when the Australian Mosquito came over that afternoon. The Mosquito made a dummy run, the engine came to a standstill, and the fireman and the driver emerged and hurried into the woods.
>
> Twice again the Mosquito attacked, using cannon and machine gun fire, and steam rose from the perforated engine boiler to a height of more than 100 feet. Panitz and Williams returned to Combourg and set course for the Montfort railway. After following the main railway, they sighted a goods train on a branch line at St Meen, and this they twice attacked with cannon and machine gun. Steam was rising from the engine when the Mosquito turned in the direction of the main railway. A goods train was found stationary outside Caulnes Station. Machine gun bullets struck the boiler, and the Mosquito moved on. A moving goods train was found on the line about two miles south-east of Dolo, and it, too, was left severely

28 Rex Timbury, interview with author, 2 Jan 1995.

29 A night-fighter variant of the FB.VI had originally been intended for production in Australia in addition to an attack version, Sharp & Bowyer, op cit, p.111. The first Mosquito to be delivered to Australia, in 1942, was a F.II model, DD664 (which became A52-1001). Later in 1943, T.III trainers were delivered to 5 OTU at Williamtown. When Australian production eventually got underway, it was mainly of the Mosquito FB.40 attack variant.

30 Wg Cdr G Panitz DFC, 404891. No 456 Sqn, 60 OTU; comd 464 Sqn 1944. Master baker of Southport, Qld; b Boonah, Qld, 21 Sep 1915. Killed in action 22 Aug 1944.

> damaged. The fifth attack was made on a train about five miles east of Lamballe, where a few minutes later, in a railway yard, three goods trains were also seen. One was hit in the firebox; a thick column of steam and smoke arose, and the train stopped.[31]

The following night, 7 May, was another success: the first kill for a 456 Mosquito. Flying Officers Geoff Oxlade and Don Shanks (HJ650) were airborne at 2310 hours, under the radio callsign "Passport 33", being controlled by Sturminster GCI, to the west of Southampton. Vectored towards the Shaftesbury area, Shanks obtained a fleeting contact on the AI at one mile range, but contact was lost as the target descended rapidly. They were then put onto a second bandit, and a visual contact was obtained on the target's lights. The lights then appeared to explode as if the aircraft had blown up. The aircraft was subsequently identified as a Ju 88A bomber of KG6, which flew into the ground near Ringwood on the Hampshire/Dorset border. Don Shanks was to recall:

> This was not elucidated at the debriefing, but was clarified next day at HQ No 10 Group when Geoff [Oxlade] was sent for by the AOC and congratulated on a fine achievement. After the first Ju 88 crashed we secured a second contact at a similar height and harried it for a similar time through the very active and vicious Portsmouth anti-aircraft barrage...This second aircraft crashed into the sea off the Isle of Wight when we were within range but unable to identify or fire because of the impenetrable cloud we were in.[32]

Three day Rangers were attempted on 11 May, but had to be aborted due to the unsuitable clear weather. Weather conditions for a day Ranger had to be so bad as to be favourable for the Mosquitos, requiring a cloud overcast sufficiently low to provide protection from enemy fighters, and aid the penetration by a single aircraft into hostile territory. Two successful sorties were flown the following day, with Panitz/Williams (HJ702) immobilising two goods trains, and Ratcliffe/Lowther (DZ308) hitting a further three. There were firm rules of engagement that the crews had to abide by, for instance, Ratcliffe had seen a passenger train near Villers Bocage, but had not attacked it. Similarly, on the night of 15 May, Warrant Officer Gate Gatenby and Flight Sergeant John "Scotty" Fraser[33] (HJ701) found several trains in the Rennes area, but these were in built up areas, and the danger of causing collateral damage prevented attack.

Night Rangers continued on 15 May, with Newell/Keating (DD789) scoring a steam train, and Panitz/Williams (DZ269) getting two electric trains and shooting up a motor transport near Rennes. The following night,

31 Johnson, op cit, p.38.

32 Donald Shanks, letter to David Vincent, 23 Apr 1981.

33 Flt Lt J M Fraser, 404897. No 456 Sqn, 62 OTU, 456 Sqn. Civil servant of Brisbane; b Clermont, Qld, 21 Jul 1919.

DZ269, 456 Squadron's top-scoring 'train-buster'.

Flight Lieutenant Bill Hyem[34] and Flight Sergeant Johnny Brayne[35] (HJ702) set an engine alight, followed on the next night by Flight Sergeants "Tony" Samson[36] and Alec Abbey (DZ269), who scored three trains around Lamballe, near the north coast of the Brest peninsula.

This night, 16 May 1943, was also to see some particularly significant intruder sorties for the Australian night flyers. It was the night of Wing Commander Guy Gibson's 617 Squadron "dambuster" raids. Two crews, Ratcliffe/Lowther (DZ308) and Oxlade/Shanks (HJ701), deployed to Castle Camps aerodrome, near Cambridge, where they briefed, as one of the navigators, Donald Shanks, recalled:

> We were briefed to fly to Holland and to orbit a night fighter airfield for an hour and a half without attacking or firing at enemy aircraft. Ratcliffe

34 Sqn Ldr R W Hyem, 402863. Nos 96 & 90 Sqns RAF, 456 Sqn, HQ 9 Gp, 54 & 51 OTUs, 464 Sqn. Jackeroo of Gunnedah, NSW; b Gunnedah 28 Feb 1919.

35 Flt Lt E J Brayne, 403715. No 456 Sqn, 54 OTU, 464 Sqn, 1 OTU, 5 OTU. Clerk of Sydney; b Sydney 9 Aug 1919

36 Sqn Ldr C S Samson, 406627. No 488 Sqn RNZAF, 456 Sqn, 63 OTU, 456 Sqn. Clerk of Freemantle, WA; b Perth 18 Jan 1919.

> and Lowther drew an airfield near the Hun border, we drew Gilze, near Rotterdam. We completed the operation, as did 26 (28 in all) other night fighters and returned to Castle Camps at 0405 hours, and then to Middle Wallop at 0600, a very long day! We were informed later that day by Wing Commander Dwyer that we had been keeping the homeward track clear for the most incredible and heroic air operation of the war, the bombing of the Mohne and Eder Dams...[37]

Ratcliffe and Lowther flew at 2,000 feet to the north of Amsterdam, over the Zuider Zee, followed by Oxlade and Shanks, who flew a route further south at 3,000 feet, below Rotterdam, and over Zeeland. They experienced heavy flak and machine gun fire over Gilze airfield, but both aircraft returned safely. This was certainly a little-known episode in the annals of the RAAF.

A further six Rangers were flown over the nights of 20 to 22 May, but without inflicting more damage to the German rail system, when the effort was switched to night intruding sorties. This intruding involved standing at readiness to patrol enemy night fighter bases in support of Bomber Command. Over 23 to 29 May, seven intruders were flown over the Dutch airfields Bergen, Venlo, Twente and St Trond, but no interceptions were made. The enemy was always prompt in dousing the airfield lights each time a Mosquito approached.

During the month, the 'B' Flight Commander, "Handlebar Hank" Richards, was posted to command another night fighter unit, No 151 Squadron RAF. His place was taken by one of 456's pilots, Bill Hyem. Squadron Leader Hyem was the first RAAF officer to become a flight commander on 456, and with "Robbie" Robinson commanding 'A' Flight, both these posts were now held by Australians.

On 30 May, the CO, Wing Commander Mickey Dwyer and Flying Officer Don Shanks (HJ701) were tasked with a morning Ranger to north-west France, but were forced to return due to unsuitable weather. As they made landfall back at Lizard Point, near Land's End at the tip of Cornwall, they lost the port engine and were unable to feather the propeller, which caused excessive drag. Not being able to maintain height on one engine, and with the good engine losing oil pressure, Dwyer was forced to crash land wheels-up in a field near Crediton, north-west of Exeter, writing-off the Mosquito. Both crew were injured, and admitted to Exeter Hospital, with Shanks being placed on the seriously ill for the next ten days. The CO had a broken leg, which necessitated his immediate replacement in command of 456, while Shanks was not to return to flying for over six months. The new CO was Wing Commander George Howden[38], an Australian serving with the RAF, who took command of 456 on 12 June.

While the intensive Ranger and intruder activity was being maintained, one of the Squadron's training aircraft, the Magister, had crashed on 23

37 Donald Shanks, letter to David Vincent, 23 Apr 1981.

38 Wg Cdr G Howden DFC, 39738 RAF. No 68 Sqn RAF; comd 456 Sqn 1943; HQ ADGB; omd 68 Sqn RAF 1944-45. Regular air force offr of Perth; b Perth 8 Aug 1914.

Sergeants' Mess at Middle Wallop Gordon Gatenby (background) with Tony Samson in front of him, (to right) Scotty Fraser, Tom Landy, Allan Keating.

May. Flying Officer Roger Hitchcock was on a joy flight with one of the RAF groundcrew, LAC H Roberts[39], when the Magister (P2464) was spun into the ground near the airfield, at Winterslow, killing both occupants.

As Rangers continued into June, 456 suffered its first loss on operations. Warrant Officer Red Ratcliffe and Flight Sergeant Ron Lowther (DZ308) failed to return from an afternoon sortie on the 2nd over France. The weather was reported as fine, and it appeared they had been caught out by enemy fighters, without the safety of cloud cover.

The extra tasking that had come to 456 Squadron showed the flexibility of the night fighter units, and indeed the multi-role capability of the Mosquito. Now new missions had arisen. The "Flower" patrol began, with Mosquitos patrolling enemy night fighter airfields while Bomber Command aircraft passed through the night fighter belt. In addition was the need for the protection of Coastal Command aircraft from the *Luftwaffe* fighters operating in the Bay of Biscay. This became the Instep patrol.

39 LAC H R Roberts RAF. Regular airman of Barmouth, Wales. Killed in flying accident 23 May 1943.

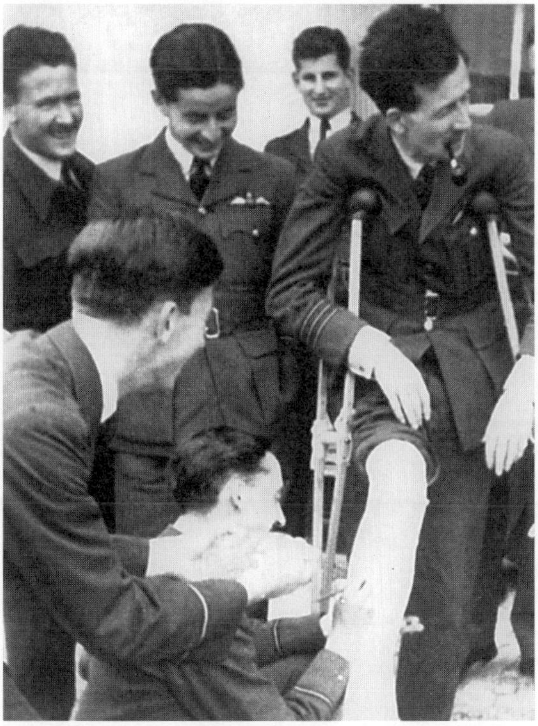

Wing Commander Mickey Dwyer returns to 'The Pheasant' for the mandatory signing of the broken leg, June 1943.

The new CO, Wing Commander George Howden DFC (right) and navigator Flight Lieutenant Dan Norris-Smith

CHAPTER 8

INSTEPS

Instep patrols had been flown by Fighter and Coastal Commands since late 1942. With German U-boats issuing from their bases in Western France to attack Allied shipping in the Atlantic, Coastal Command aircraft were hunting the U-boats. In turn, the *Luftwaffe* then sent out Ju 88 and Bf 110 fighters to attack the lumbering Coastal Command aircraft. To counter this new air threat, the RAF had devised a new role, long-range day fighter patrols of the Western Approaches and the Bay of Biscay – the Instep patrol. In one engagement on 8 October 1942, the RAAF Spitfire unit in Fighter Command, No 453 Squadron, had destroyed at least five Bf 110 *Zerstörers* ("Destroyers") engaged in these operations.[1] By the following year, the vulnerable Bf 110 had been withdrawn from this role, and it was now the Ju 88 fighters of KG40, from bases at Merignac and Cognac, that were interfering with the Sunderland U-boat operations. To extend the range of the Instep patrols, from May 1943 Mosquitos were used as long-range fighters, and Predannack airfield, to the south-west on the top of the cliffs at the tip of Cornwall, was used as an advanced landing ground.

On 4 June, No 10 Group advised that three 456 Squadron crews were required immediately for Instep patrols from Predannack, to operate under the orders of Coastal Command in a composite squadron, made up from detachments from night fighter squadrons. The first crews were deployed immediately to Predannack, and under the lead crew, Flight Lieutenant Peter Panitz and Flying Officer Dickie Williams, were Flying Officer Johnny Newell and Flight Sergeant Allan Keating, with Flight Sergeant Bob

1 J W Bennett, *Defeat To Victory*, RAAF Museum, Point Cook, Vic, 1994, pp.100-2.

Mosquitos set off

Richardson[2] and Sergeant Tommy Landy[3]. The Mosquito IIs were fitted with long-range tanks to enable five hour sorties, and the first of these marathon trips were flown in company with two No 25 Squadron Mosquitos on 7 June. The Insteps were flown at very low level in a wide "battle" formation, normally with four aircraft in formation split into two elements a mile abreast, with each aircraft in the pair swept by about 800 yards. This enabled a good lookout to be maintained, searching the sky for any Ju 88s, as the formation crossed the Bay of Biscay, almost as far as the Spanish coast.

The first aerial action for the Australians in the Bay occurred on 11 June. A patrol of six Mosquitos, three each from 25 and 456 Squadrons, set course from Lizard Point at 1500 hours, but Panitz (HJ702) had engine trouble and had to turn back. The others pressed on at zero feet and, after almost an hour and a half, five Ju 88Cs were spotted, flying at about 6,000 feet. The Mosquitos climbed towards the Germans, who were flying in line astern in a port orbit. The formation leader, Flight Lieutenant Joe Singleton[4] (DD757) of 25 Squadron, shot down one aircraft (coded FB+HZ of KG40), and Flying Officer Newell and Flight Sergeant Keating (DZ269) claimed two as damaged. The following day, the three 456 crews flew an uneventful Instep with an aircraft from No 410 Squadron RCAF.

Intruders and night Rangers continued, and another Instep was flown on 19 June, when Warrant Officer Gordon Gatenby and Flight Sergeant Scotty

2 Flt Lt R W Richardson, 407849. No 600 Sqn RAF, 456 Sqn, 51 OTU, 456 Sqn. Draughtsman of Adelaide; b Rangoon, Burma, 18 Apr 1916.

3 Flt Lt T M Landy, 404706. No 456 Sqn. Clerk of Brisbane; b Brisbane 12 Sep 1918.

4 Wg Cdr J Singleton DSO DFC AFC RAF, 69431. 60 OTU, No 25 Sqn RAF, TRE Defford, 25 Sqn, HQ Ftr Cd, CFE; comd 25 Sqn 1947-48; HQ 12 Gp, HQ 2 TAF. Regular air force offr; b Leyland, Lancs, 12 Mar 1916. His observer was Flg Off Geoff Haslam, and the other two RAF aircraft on this patrol were flown by Flg Off Jimmy Wooton/Plt Off J M Dymock (DZ685) and Flg Off Jack Cheney/Plt Off J K Mycock (DZ688). Both failed to return from an Instep two days later, together with a 410 Sqn aircraft, Flg Off R B Harris/Sgt E H Skeel (DZ753), all believed to have been shot down by Focke Wulf Fw 190s.

Gordon 'Gate' Gatenby and John 'Scotty' Fraser

Fraser (DZ299) operated with three Mosquitos of No 151 Squadron. After a long transit at sea level as far south as Cape Ortegal, on Spain's north-western tip, Gatenby saw barrage balloons in the distance. Upon calling this to the formation, he saw the "balloons" moving, and realised it was eight Ju 88s circling a trawler. The Mosquitos pulled up and made a head-on attack to break up the German fighters, and Gatenby saw his strikes hitting home on his target. Descending back down through cloud, Gatenby found the other aircraft had disappeared, and after circling the area, the four Mosquitos were eventually able to rejoin for their return to base.[5] One Ju 88 had been shot down by the formation leader, Squadron Leader Henry Bodien[6], and Gatenby had damaged another. In both of these engagements, it was the leader who was able to able to get a good shot, while Nos 2 and 3 could just get an occasional squirt as they flashed through, dispersing the Ju 88 formation.

On Gatenby's next Instep two days later, again flown in company with 151 Squadron, a U-boat was spotted on the surface. The Mosquitos attacked, and strikes were observed all around the conning tower. The U-462 was forced to return to port for repairs. This was the only day that No 456 Night Fighter Squadron could claim to have damaged a submarine. Another unusual occurrence was on 23 June, when Flying Officers George Houston[7] and Lars "Gus" Engberg[8] (DZ297) were on an air sea rescue (ASR) patrol. They spotted a Spitfire pilot down in the sea, and skilfully dropped a dinghy beside him. This was no easy feat in a Mosquito, requiring Engberg to jettison the crew door and push out one of the crew's dinghies. The dinghy landed well within reach of the exhausted pilot, who was able

5 Gordon Gatenby, interview with author, 17 Dec 1994.

6 Sqn Ldr H E Bodien DFC RAF, 45720. No 151 Sqn RAF.

7 Flt Lt G R Houston, 415146. No 456 Sqn. Petrol depot keeper of Geraldon, WA; b Geraldon 16 Jul 1921.

8 Flt Lt L G Engberg, 401642. No 256 Sqn RAF, 456 Sqn, 5 OTU. Woolbuyer of Melbourne; b Stockholm, Sweden, 1 Apr 1913. Engberg had previously served in the Swedish army before migrating to Australia.

to enter it, to be later picked up by a Walrus amphibian. Three nights later, on 26 June, Flight Sergeant Bob Richardson and Sergeant Tom Landy (HJ943) flew another successful ASR patrol, being scrambled to search for a Spitfire pilot who had baled out in the mid-Channel. Guided by flares dropped by other aircraft, they orbited and saw a torch flashing from the sea. Good fixes were obtained from their radio transmissions, and rescue launches were able to be directed to pick up the downed pilot.

During June, No 456 Squadron had recorded the highest flying hours of any unit in 10 Group, achieving this milestone without any flying accidents, and by July, the Squadron was operating from three bases on its variety of missions. Intruders and Rangers were being flown from a forward base at Castle Camps, near Cambridge, covering enemy aerodromes in Holland in support of Bomber Command attacks. The Predannack detachment continued on Insteps, while the main body of the Squadron remained at Middle Wallop maintaining the routine night fighter state. Arrangements were being made to equip the Squadron's Ranger aircraft for bombing, and at the same time the ROs were instructed in the use of "Gee", a new grid navigation device which would be particularly valuable for Ranger work.[9]

A change to the Rangers began on 12 July, when two aircraft were despatched in company to carry out an attack on a selected target in France, followed by a hunt for trains on the way out. The Ranger veterans, Panitz and Williams (DD751), were accompanied by Flight Lieutenant Bas Howard[10] and Flying Officer Jack Ross (DZ299) for a midnight attack on a transformer station at Guerledan, on the Brest peninsula. In the face of moderate flak, Panitz attacked the transformer station, while Howard struck the nearby power house. Panitz then attacked two goods trains, which he left in flames, on his return along the Guingamp-Chateaulerun railway. The following evening, solo Rangers were carried out against rail targets by Flying Officers Ron Pratt[11] and Stewy Smith[12] (HJ939) and Wing Commander Howden and Flying Officer Jimmy Thomson (DZ269), and although both aircraft flew their complete routes, no trains were found.

Over the following week, six new fighter-bomber model Mosquitos were delivered to the Australians, these being the Mark VI variant, obtained from No 151 Squadron at Colerne. Operational experience of NF.II in the day fighter and intruder roles had led to development of the Mosquito FB.VI, which had entered squadron service with the RAF in early 1943. The FB.VI had no AI radar, but was better equipped for the Ranger role with long-range fuel tanks and bombing equipment. It had a strengthened wing for external loads, which was to become the "basic" Mosquito wing, and

9 See Appendix 10 for an explanation of Gee.

10 Wg Cdr B Howard DFC, 404787. Nos 256 & 96 Sqns RAF, 456 Sqn, 63 OTU; comd 456 Sqn 1944-45. Petrol representative of Roma, Qld; b Sydney 6 Jul 1915. Killed in flying accident 29 May 1945.

11 Flt Lt R G Pratt, 266160. No 456 Sqn, 51 OTU, 456 Sqn, HQ Ftr Cd. Grocer of Sydney; b Sydney 15 Jan 1918.

12 Flt Lt S D P Smith, 414280. No 456 Sqn, 51 OTU, 456 Sqn. Engineer of Brisbane; b 2 Jun 1913.

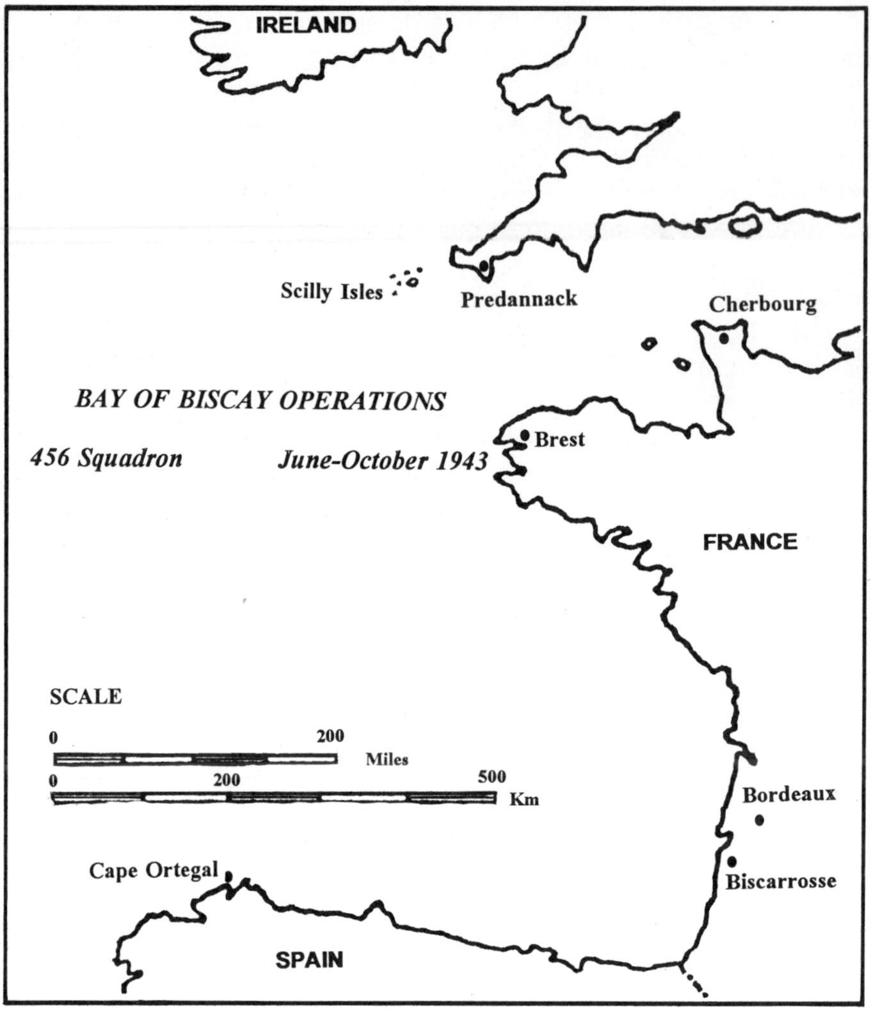

adding to the cannon and machine gun armament, it had the capability of carrying a 500-lb bombs under each wing.

The success of Mosquito Rangers and arrival of the new aircraft now enabled 456 Squadron to form a specialised Ranger flight – 'C' Flight – on 22 July. Six constituted crews were selected to form the Ranger flight, under Peter Panitz, the 'C' Flight Commander, operating under the radio callsign "Didus":

Didus 51	Flt Lt G Panitz	Flg Off R S Williams
Didus 52	Flg Off J W Newell	FSgt A J Keating
Didus 53	Flg Off R G Pratt	Flg Off S D P Smith
Didus 54	Plt Off C S Samson	Plt Off A M Abbey
Didus 55	WOff G F Gatenby	FSgt J M Fraser
Didus 56	WOff A S McEvoy[13]	Flg Off M N Austin

The members of Ranger flight soon became known as the "offensive dozen".

13 Flt Lt A S McEvoy, 411157. No 456 Sqn, 5OTU. Warehouse assistant of Sydney; b Sydney 25 Jul 1915. Killed in flying accident 19 Sep 1945.

Navigator, Sergeant Tom Landy

Rangers with the new FB.VI aircraft commenced on the night of 26 July, with a pair flown by Panitz/Williams (HJ819) and Pratt/Smith (HJ817) attacking a power station at Marais de St Miche, near Brest. Making landfall at Ile de Batz, both aircraft strafed the target, causing explosions and leaving the power station in flames. The following night it was the turn of Samson/Abbey (HJ817) and Gatenby/Fraser (HJ818), attacking the transformer at Plancoet, near Dinan. Continuing his sortie to the south, Gatenby attacked a train, which pulled up to hide in a cutting, north of Broons. After his first pass, he was being fired on by the enemy troops that had detrained, but made several more determined attacks. His aircraft was hit in the fuel tank in the wing, but as it was self-sealing, it presented no problem to his safe return to base.[14] The other members of the Ranger flight flew intruding missions of airfields in Belgium and Germany until their recall to Instep sorties.

Instep activity had been maintained by other units during July with Nos 264, 307 and 604 Squadrons flying from Predannack, and No 25 Squadron also forming a separate Ranger flight of FB.VIs. The call came on 2 August for 456 to reinforce the Predannack operations, and the Ranger flight was sent. It was to remain employed there for the next two months.

Panitz and Williams (HJ817) led a patrol that first evening, but their only excitement was when they were fired on by a couple of Royal Navy destroyers. Although no damage resulted, when the next evening's patrol found the friendly naval units, the Mosquitos were a little more wary as they approached. Again it was the norm to fly formations of four aircraft, but

14 Gordon Gatenby, interview with author, 17 Dec 1994.

Mosquito FB.VI cockpit.
The starboard tunnel to the nav/bombardier's position in the nose was blanked off in the fighter and fighter-bomber variants.

now the Ranger flight was six strong, it was possible to fly wholly 456 constituted patrols, without mixing the formations with other squadrons. The patrols would normally be flown south-west from the Scillies, at zero feet, below the German radar cover from the Brest peninsula, in an open formation of four aircraft in line abreast. Once far enough south in the patrol area, it was usual to climb and fly at about 200 feet to ease the strain on the pilots of flying at deck level. The lead observer did the navigation for the formation, by checking the wind on the sea surface and taking an occasional Gee fix, but it was of vital importance to have all available eyes scanning the horizon in every direction.

The first action was when the Mosquitos attacked an armed trawler on 7 August, off La Rochelle, with one aircraft, flown by Samson/Abbey (HJ816), being hit in the wing by anti-aircraft fire. The fighters blew away the superstructure of the bridge, and left the vessel burning. During the latter half of August, the detachment flew six Instep patrols of four aircraft on each occasion, but there were no further engagements.

Meanwhile, on 14 August, the order was received for 456 Squadron to move to Colerne in three days time, to change places with No 151 Squadron. This unit's Mosquitos were now better equipped with the latest AI Mark VIII, and more likely to succeed in low-level interceptions along the south coast. Accordingly, the main party moved by road convoy from Middle Wallop to Colerne on 17 August. Colerne airfield, in Wiltshire, was on top of a hill overlooking the historic city of Bath. The officers lived about four kilometres away, at Ashwick Hall, a huge baronial residence set in a large park.

Rangers were flown over the Brest peninsula in September, with several trains being damaged. The greatest success occurred for the Ranger flight on 21 September, when six Ju 88s were spotted over the Bay. Just below the cloud cover, the enemy were flying in line astern when the Mosquitos attacked. Panitz and Williams (HJ818) closed in on the last aircraft, which hesitated in evading, and firing from 200 yards, Panitz saw hits on the cockpit and starboard engine, which started smoking. The Ju 88's tail gunner had now stopped firing, as Panitz shot another burst into the starboard engine, sending the aircraft diving down. Samson and Abbey (HJ816) then fired at it from close range, causing the port engine and fuel tank to burst into flames, and the Ju 88 crashed into the sea.

Meanwhile, as the German formation had split into two groups of three, Gatenby and Fraser (HJ814) had gone for the higher section. They attacked an aircraft flying in a fine layer of cloud at around 5,000 feet, delivering four bursts from dead astern. Strikes were seen all over the German aircraft, and it was seen covered in black smoke from wingtip to wingtip as it dived into thick cloud.[15] Newell and Keating (HJ815) attacked one Ju 88 flying at 7,000 feet, from 600 yards. It entered cloud with smoke pouring from its starboard engine. They then attacked another aircraft, without seeing any hits.

15 Gordon Gatenby, interview with author, 17 Dec 1994.

Ten of the "Offensive Dozen"
Left to right: Flg Off M N Austin, WOff A S McEvoy, Flg Off R S Williams, Flt Lt G Panitz, Plt Off G Gatenby, Plt Off J M Fraser, Flg Off J W Newell, Plt Off A M Abbey, Flg Off S D P Smith, FSgt A J Keating (front).
Missing from photo were Flg Off R G Pratt and Plt Off C S Samson.

The Predannack detachment.
Not all the ground crew can be identified, but Cpl A Coates and Sgt Lance Sutherland are in the centre row, fifth and sixth from the left. The aircrew are: Plt Off Tony Samson (centre, far left), Flg Off Dickie Williams (centre, far right), (front row, from left) Plt Off Alec Abbey, Flg Off John Newell, FSgt Allan Keating, Flt Lt Gordon Panitz, Plt Off Scotty Fraser, Flg Off Stewy Smith, WOff Arthur McEvoy, Flg Off Merv Austin.

The total claims for the engagement were one destroyed (shared by Panitz and Samson), one probable to Gatenby, and one damaged to Newell. A dinghy was later seen in the water, perhaps from Gatenby's "probable".

An Instep brought more success on 3 October. Panitz and Williams (HJ817) led out an evening patrol, consisting of Newell/Keating (HJ815), Pratt/Smith (HJ819) and Samson/Abbey (HJ814), to attack the seaplane base near Lake Biscarrosse, south-west of Bordeaux in the Gironde. They were met by intense flak, but were disappointed to find there were no enemy aircraft on the lake, only launches and patrol boats. Panitz sank a refuelling launch and Pratt damaged a tender. Newell scored hits on a flak barge, and the others strafed the slipway. Having set course for the coast, a minesweeper was seen about five kilometres south of Biscarrosse, which opened fire on the Mosquitos. Panitz attacked the vessel, and on setting it alight, he immediately spotted a Ju 88 flying at about 200 feet above the water. He closed to 75 yards and fired a 2-second burst. The enemy aircraft blew up and hit the sea in flames.

One aspect of the "Bay" operations had been the problems caused to engines. On this patrol, Panitz's engine had cut as he crossed the French coast on the way in, and had continued to do so for the rest of the trip. In addition, Newell's port engine had cut over the target, with his starboard cutting intermittently all the way home, while Pratt's starboard engine had also cut out over the target in the midst of the most intensive flak. At extremely low level, flying below 50 feet over the sea, the salt air was being sucked into the fuel breather tubes, and on these excessively long sorties, this would cause the salt to clag up, stopping fuel from entering the carburettor.[16] Fortunately, no 456 Mosquitos were lost to this problem.

While operations had been conducted from Predannack, an accident had occurred back at Colerne on 29 September. Flight Lieutenant Ern Griffith[17] was air testing a Mosquito (DZ690), and had taken one of the Squadron's mechanics, Corporal Bill Blakeley[18], for a joy flight:

> ...one of the groundcrew chaps, Bill Blakeley, after working all night on the undercarriage mechanism, decided to go on a test flight...the pilot, after testing the undercarriage, came over the 'drome and did a roll at approximately 1,000 feet and one engine fell out, consequently they went into the deck, and both killed.19

It was a cruel twist of fate that such a freak accident should occur in sight of the personnel of the Squadron. The starboard engine had indeed separated from the wing, ripping away the engine fairings, and still with propeller attached, had buried itself in the airfield at Colerne. The Mosquito

16 Gordon Gatenby, interview with author, 17 Dec 1994.

17 Flt Lt E H Griffith, 403334. No 96 Sqn RAF, 456 Sqn. Of Narrandera, NSW; b Albury, NSW, 22 May 1916. Killed in flying accident 29 Sep 1943.

18 Cpl W H Blakeley, 26160. No 456 Sqn. Of Adelaide; b Adelaide 15 Oct 1919. Killed in flying accident 29 Sep 1943.

19 Ron Vidler, letter to author, 5 Dec 1994.

On a Mossie at Colerne – from left: Flg Off Col Griffin, Sgt 'Ali' Barber, Flg Off Harry Smith, Flg Off Jock Donald, Cpl Reg Jeans, Flt Lt Hoppy Williams, Cpl Tony Crooks and Flg Off Doug Wiley.

then came down, almost flat, crashing only 20 metres from a hangar, and exploding on impact. The efforts of the fire tender crew, on the spot immediately, proved unavailing. An investigation subsequently found that the engine mounting bolts had stripped, necessitating all others be changed.[20]

No 456 Squadron was advised on 6 October that the Mosquito FB.VI aircraft were to be handed over to the OTU at RAF High Ercall, and so the detachment at Predannack ceased immediately. With one of the flight commanders, Squadron Leader Bill Hyem, being posted to 9 Group HQ, Peter Panitz was promoted to Squadron Leader to take charge of 'B' Flight.

Over the second half of 1943, enemy raids over England had been scarce. For instance, No 151 Squadron, flying from Middle Wallop and Colerne, had only one victory, while No 125 Squadron only had one engagement. All of 456 Squadron's claims had been from the Instep patrols. The intruder and Ranger offensive had grown to a higher priority, from little activity in the first half of 1943, to now becoming an almost nightly occurrence. It was now the role of these fighter squadrons to continue through the night, on a relay system, to extract the greatest possible nuisance value against rolling stock on the move, marshalling yards, and enemy airfields waiting for returning bombers.[21]

The Ranger leaders on 456, Panitz and Williams, were both awarded the DFC on 29 October 1943, and the citation to Panitz's award gave the pair's tally as 13 trains. Wing Commander Peter Panitz was to take command of another RAAF Mosquito unit, No 464 Squadron, a month after the D-Day invasion, leading that Squadron on operations for the first time on 3 July.

Panitz, with his customary navigator Dickie Williams, did not return from an attack on rail communications in central France on 22 August 1944.

20 Ron Vidler, telephone interview with author, 9 Dec 1994; Len Stamp, letter to author, 11 Dec 1994.

21 ACdre L E O Charlton, *Britain At War – The Royal Air Force, April 1942-June 1943*, Hutchinson, London, 1945, p.50.

Anyone for tennis?
A group of the ground crew at Ashwick Hall, the officers' quarters at Colerne, August 1943. From Left: Eric Soames, Ben Davis, Jack Beatty, Bill Clapperton, unknown, Lyle Hale.

Squadron officers, September 1943. Sqn Ldr Bill Hyem, Flg Off Harry Smith, Wg Cdr George Howden, Flt Lt Dan Norris-Smith and Flt Lt 'Peter' Panitz.

Back Row, left to right: FSgt Ivor Sanderson, Sgt Charles Nicholas RAF, FSgt Dick Harris RAF, FSgt George Hough RAF, FSgt Stan Williams, WOff Bob Wilmoth, Flg Off George Houston, Flg Off 'Gus' Engberg, Flg Off J White RAF, WOff Johnny Brayne, Flg Off J Smith RAF, FSgt 'Butch' Hodgen
Centre Row (L-R): Plt Off Tommy Landy, Plt Off Andy Kellett, WOff R Forbes RAF, Sgt Nelson RAF, Flg Off Doug Wiley, Flg Off Harry Smith, Flt Lt 'Peter' Panitz, Flg Off Stewy Smith, Plt Off Keith Roediger, Flg Off Bob Dobson, FSgt Tommy May, Sgt A McCormick RAF
Front Row (L-R): Plt Off Bob Richardson, Flt Lt Bob Lewis, Flg Off Clive Brooks RAF, Flg Off Bill Donald RAF, Sqn Ldr Geoff Robinson RAF ('A' Flt Cdr), Wg Cdr Geoff Howden RAF (CO), Sqn Ldr Bill Hyem ('B' Flt Cdr), Flt Lt Dan Norris-Smith, Flg Off 'Paddy' Bailey RAF, Flg Off P Biggin

'Peter' Panitz DFC

HONOURS AND AWARDS[22]

DISTINGUISHED FLYING CROSS

FLIGHT LIEUTENANT GORDON PANITZ

CITATION

Flight Lieutenant Panitz has taken part in many operational sorties. In operations during the last 4 months he has damaged 13 trains and several enemy electrical installations. He has at all times displayed fine leadership, skill and devotion to duty..

FLYING OFFICER RICHARD WILLIAMS

CITATION

Flying Officer Williams has flown as navigator on a number of operational missions. His work has been faultless and has contributed much to the success of these sorties. He has assisted in damaging many trains and transport in enemy territory and has participated in raids on transformer and power stations which have proved most effective. This officer has invariably displayed great courage, dash and determination.

22 London Gazette No 36226, 29 Oct 1943.

CHAPTER NINE

CENTIMETRIC RADAR

One of the most significant technological leaps of the Second World War was the development of centimetric radar.The breakthrough of radar operating on a wavelength below 10cm had become possible by the British-invented cavity magnetron valve. Being capable of generating powerful transmissions at short wavelengths (microwaves), centimetric radars were able to overcome the main problem that had plagued the Mk IV generation of AI sets. These powerful transmissions were beamed by a directional antenna, thereby largely eliminating target blips being lost in the ground returns. British development of centimetric radar was an example of technology where one side was to build up an overwhelming lead.[1]

AI Mark VII

The first 10cm microwave wavelength AI radar, which operated on a frequency of 3000 MHz (ie 3 GHz), was developed in England by GEC and had first flown during November 1941 in a Beaufighter. 125 hand-made Mk VII sets were produced for fitting to Beaufighters, and became operational by June 1942 with No 604 Squadron RAF.[2]

The AI Mk VII had greatly improved performance by employing a spinning-dish scanner, which was a dipole aerial, fixed to the aircraft and always pointing ahead, with a reflective dish (scanner) which rotated in expanding circles about this fixed axis. This spirally rotating antenna produced a narrow beam of about 12 degrees straight ahead and, by electronic switching, the same aerials could be used for transmission and

1 Jones, op cit, p.129.

2 Robertson, op cit, p.35.

reception. The transmitter and receiver aerials were under a perspex dome on the nose, known as the "thimble" nose.

Apart from the absence of ground returns, for the operator the other big difference was the display was presented on one scope only. The CRT was about six inches in diameter, with the centre of the display representing the aircraft, which gave both range and angle off simultaneously. A target within 45 degrees of the nose could be displayed, showing as a circle, or part of an arc, of light. A target dead ahead and co-altitude would be shown as a full circle of light, the radius of the circle representing the target range. If the target was 10 degrees to one side, half a semi-circle on that side would be shown. Beyond that, the arc got smaller, out to the 45 degree of boresight limit. Elevation was shown in a similar fashion.

The AI Mk VII had a maximum range 5 to 6 miles, even from low level, because of the beamed transmissions.[3] Experienced radar operators could get a 7 mile range. With the introduction of the fully productionised AI Mk VIII, the Mk VII was relegated to training in late 1943, its short time in service indicating the tremendous advances occurring in radar development, with a generation of AI radar barely lasting two years. The British had worked in collaboration with the Americans with AI Mk VII in designing workable radar circuits, and this led to the US-developed SCR-520 10cm AI radar, effectively the beginning of American air-to-air radar systems.

AI Mark VIII

The AI Mk VIII was the mass-produced fully-engineered version of AI Mk VII, with the RAF designation ARI 15093. It was basically the same as Mk VII, but with a beacon-homing facility and capable of IFF interrogation. It had the advantages of a greater maximum range and lower minimum range than AI Mk VII, capable of operating out to 7 or 8 miles.[4] As with the Mk VII, the beam width was about 12 degrees straight ahead, but broadened to 30 degrees at the 45 degree limits. On these limits, target detection was reduced from 7 miles to 2 miles, and relative movement was difficult to assess.

By early 1943, twelve squadrons of the RAF night fighting force were flying Beaufighters, some equipped with the latest Mk VIII centimetric radar sets, along with six operational units of radar-equipped Mosquitos.[5] The AI Mk VIII "thimble" type nose radome replaced the four machine guns of the Mosquito NF.II, and this became the Mosquito NF.XII. This model, first flown in August 1942, was delivered to No 85 Squadron and became operational in April 1943. When the long-range Mosquito FB.VI airframe was produced with this AI, the version was designated Mosquito NF.XIII, which flew in August 1943. By that October, the Beaufighter had standardised on AI Mk VIII, and by the end of 1944, AI Mk VIII had become standard for RAF night flying squadrons in the Mediterranean theatre.

3 Howard-Williams, op cit, p.175.

4 ibid, p.175.

5 Bowyer, op cit, p.150.

Mosquito NF.XIII with the thimble nose radome housing the AI Mark VIII.

As with other radar developments previously, crews in AI Mk VIII-equipped aircraft were forbidden to fly over enemy territory, and so the task of intruding was left to the units equipped with the older AI Mk IV equipment, such as 456 Squadron. It would not be until D-Day that the "state of the art" AI Mk VIII equipment would be risked over hostile territory. In all, some 2,500 Mk VIII were produced in England by GEC and Ekco.[6]

AI Mark IX

The AI Mk IX was a considerable redesign of the Mk VIII, developed over 1942. The narrow spiral scanner could be controlled by the RO to point exactly at the target, slewing and elevating the antenna until a small ring encircled the blip. By triggering a thumb-switch, the set could "lock on" to the target and track automatically. The display of this sophisticated AI could also be projected on to the fighter's windshield, to give a "head-up display" to the pilot of the radar picture, the artificial horizon and the gunsight. There were problems, however, as the display was too dim by day, too bright by night, and was not focused at infinity.

6 Gunston, op cit, p.69.

AI Mark IX was first flown in a Beaufighter during November 1942.

The prototype aircraft was tragically shot down by an eager Spitfire pilot on 23 December, and then development of the system was suspended in early 1943 in favour of the imported AI Mk X. Limited production did eventually begin, but the Mark IX did not reach squadron service before the war ended. The system replaced the Mark X AI in the Mosquito NF.38 during 1947, to ensure radar development did not rely solely on American sources of supply.

AI Mark X

The Americans improved their SCR-520 AI into the SCR-720, which was first delivered to the USAAF by Western Electric in November 1942. Operating on 10cm wavelength, the first SCR-720 arrived in Britain in November 1942, and flew on board a Wellington during December. After successful trialling on a Mosquito NF.II during 1943, the SCR-720 was then supplied under Lend/Lease, to become the new standard AI for the RAF, as the AI Mk X. The first Mosquito variant to be fitted with AI Mk X was the basic NF.II, which became the Mosquito NF.XVII, fitted with the bulbous "universal" radome, which first flew in March 1943. Becoming operational with 85 Squadron RAF that November, the Mosquito NF.XVII equipped 456 Squadron from January 1944, and was the RAAF's introduction to centimetric AI radar.

In addition, the Mosquito NF.XIX (equivalent of NF.XIII but with Mark X AI under the universal radome) was placed in production, and flew in April

Mosquito variants that were fitted with the universal nose could accept either the AI Mk VIII or Mk X, however, the former radar was not introduced to 456.

1944. A variant of this model, with more powerful two-stage Merlin engines, was the Mosquito NF.30, which retained the AI Mk X, and equipped 456 Squadron from December 1944.

The Mk X reverted to two scopes instead of one, providing the RO with two rectangular tubes, one showing a plan view, the other an elevation display. The right CRT, known as a "B-scope", showed range and bearing (angle off the nose) out to 10 miles, being selectable on a 10 or 5 mile scope. The RO then transferred his attention to the left scope, which was a vertical clock face in front of the aircraft, known as the "C-scope". This elevation display represented angle off in azimuth and elevation. With many more knobs for adjustment, the AI Mk X was more difficult to use than Mark VIII, but these selections could mainly be preset during the NFT.

The aerial and the reflective scanner rotated through a full 360 degrees horizontally, but only transmitted in a 60 degree arc in the forward sector, and the assembly could nod in elevation from 50 degrees up through to 20 degrees down. The scanner nodded as it rotated, to search above and below the aircraft's flight line. A constant 10 degree beam width enabled the maximum range of 10 miles to be achieved from side to side (typically a 5 mile range could be achieved from a height of only 500 feet), and in addition, the set was less susceptible to clutter from ground returns.[7]

Despite the broad search area of AI Mk X, this was only forward of the aircraft, and there was no warning of the fighter itself being hunted. This led

7 Meadows, Part 4, *Aeroplane Monthly*, Jan 1993, p.43; Howard-Williams, op cit, p.175; Gunston, op cit, p.81.

AI Mark X display in a Mosquito NF.30, with a Monica CRT display underneath.

The two displays of the AI Mark X, the left being the C-scope, representing a vertical clock face, and the B-scope, right, indicating range and bearing on a selectable 5 or 10 mile scale.

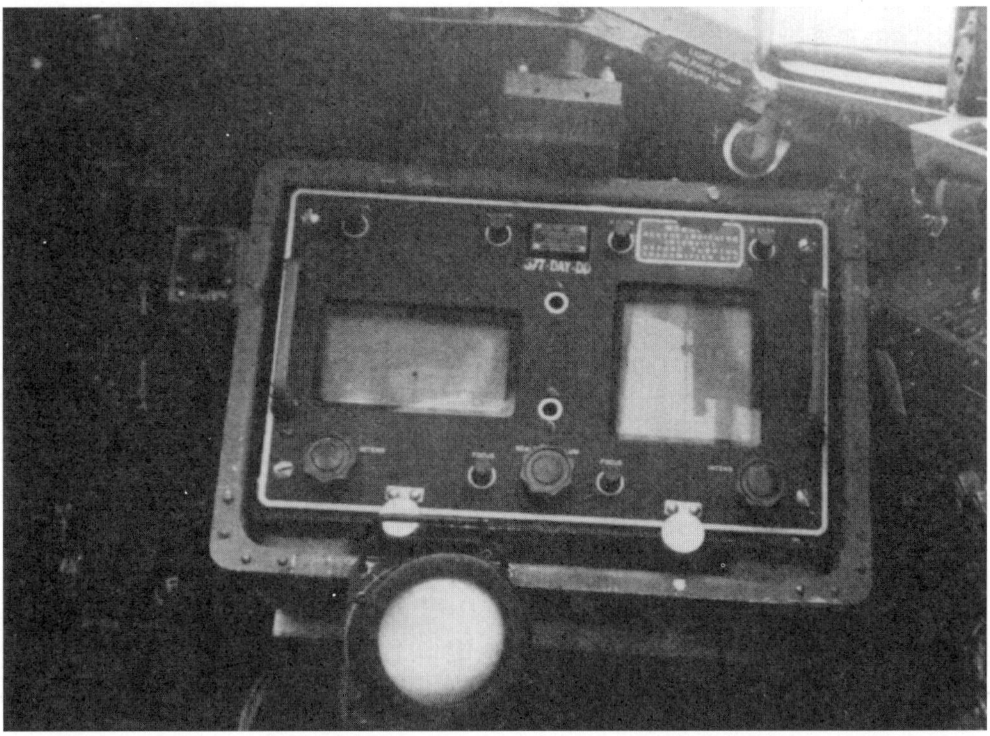

to tail warning devices, such as "Monica", which became standard equipment for AI Mk X-equipped aircraft. AI Mk X aircraft were also equipped with the SCR-729 radar receiving beacons, which proved an excellent facility for homings to base. The AI Mk X proved reliable, giving good forward coverage which was largely free of ground clutter. This system became standard equipment for nearly all British and US night fighter squadrons. Having been introduced into service in 1943, the AI Mk X became the RAF's main night fighter radar for the next 13 years.

Later AI Development

The other main centimetric AI radar developments in UK during the war were the AI Mk XIV and the Mk XV. The AI Mk XIV, also known by the Royal Navy as AIA, was flight trialled over late 1944 and early 1945, but was not introduced to service. The AI Mk XV was the American 3cm wavelength AN/APS-4, which was flight trialled by the RAF's 100 Group from June 1944, and was adopted by the Royal Navy Fleet Air Arm. Known as ASH by the RN, the set was fitted to Fairey Firefly aircraft, but resulted in poor performance. The set was also operated by three RAF Mosquito squadrons from 1944, owing to the AI Mk X being in short supply, and proved more efficient at low level than Mk X.[8] As it had originated as an air to surface vessel (ASV) radar, AI Mk XV could also be used as a ground mapping radar for navigation.

The British had also developed centimetric AI into a ground mapping radar for Bomber Command. In April 1942, EMI modified the AI Mk VII into the H2S Mk I, which operated on 9cm wavelength, and indicated out to a range of 100 miles. By adapting the set to a PPI display for ground mapping navigation and bombing, the bombers were able to find and attack their targets in all weather. H2S began operations over Germany in January 1943, and when an RAF bomber was shot down in March, it was feared the secret of centimetric radar was out. Fortunately, the Germans were not to stumble upon this development for several more months.

8 Robertson, op cit, p.38.

CHAPTER TEN

MAHMOUD

No 456 Squadron had remained at Colerne for three months, until being replaced by No 151 Squadron, and had moved to its new base at Fairwood Common on 17 November (to replace No 264 Squadron). RAF Fairwood Common, which was six kilometres to the west of what was then the town of Swansea in Glamorgan, had been opened as a 10 Group fighter station in June 1941 for the defence of Plymouth, and was a developed airfield with three hard runways and a control tower. It was here the Australians were briefed on the new task being assigned to the night fighter squadrons, the "Mahmoud" patrol.

Mahmoud high-level bomber support operations had the aim of reducing Bomber Command's casualties. Commencing in the middle of August 1943, these offensive patrols were flown by single Mosquitos. As the new centimetric AI, the Mark VIII, was not allowed over enemy territory, the ban on AI Mark IV being flown over hostile territory was lifted, and Mahmoud sorties were flown by the older Mosquito NF.II. The Mahmoud patrols flew along the bomber streams or against the *Luftwaffe* night fighter assembly points, of which 22 had so far been identified, in the hope of intercepting enemy aircraft.[1]

No 25 Squadron RAF had carried out Mahmoud trials with Mosquitos fitted with AI Mark IV and "Monica" tail warning radar. The technique developed was for the Mosquito to fly a steady course until a contact appeared on the tail warning Monica set. The unidentified aircraft was then allowed to approach to 5,000 to 6,000 feet (as the range of any attacker was displayed on the RO's Monica indicator), at which point the Mahmoud fighter would

1 M Streetly, *Confound and Destroy*, Macdonald and Janes, London, 1978, p.28.

execute a 360 degree turn to position itself onto the enemy's tail. Hopefully the Mosquito would then have "turned the table" and no longer be the prey, but end up on the hunter's tail.

To assist the long-range night fighter protection of the bombers, No 100 (Special Duties) Group was formed on 23 November 1943 to operate radio countermeasures (RCM). The roles of 100 Group were the direct support to night bombing by attacks on enemy night fighter aircraft in the air or on the ground, employing RCM equipment to deceive or jam enemy radio navigation aids, radar systems and communications, and the collection of intelligence from all enemy radio and radar transmissions.[2] In addition to Monica equipment, 100 Group aircraft had a variety of "black boxes" in their armoury.

One such piece of "kit" was the "Mandrel" jammer, which swamped the German early warning *Freya* radar scopes, enabling the bombers to reach their targets undetected by enemy night fighters. The enemy were only then alerted to the bombers' presence by the explosion of their bombs.[3] The British boffins had also solved the problem of identifying hostile aircraft by another black box, the "Serrate", so-called because of the herring-bone pattern on the scope showing the presence of a night fighter had a serrated edge.[4] Serrate homed on the frequency of the German night fighters' *Lichtenstein* AI sets, and when within the range of the AI Mark IV, the interception was able to continue in the normal way.

In addition there was "Perfectos", one of the cleverest electronic gadgets to come out of the war, which sent out radar pulses to challenge enemy IFF sets. This enabled the intruding night fighters to trigger a response from the *Luftwaffe's* IFF sets, and provided the bearing, distance and positive identification that the bogey was indeed hostile.[5] A homing could then be made on the received reply. This led to widespread abandonment of IFF by Germans, by forcing the German night fighters to switch off their recognition sets, and creating problems for their control organisation by complicating the tasks of their ground controllers.[6]

Monica was also installed in every bomber to detect any night fighters approaching from the rear. Being an active radar, it proved to be much more effective at giving away the bomber's position to the fighter, than vice-versa. A Ju 88 night fighter, with the latest German AI equipment, that

2 Streetly, op cit, p.35.

3 Jamming of the Freyas had begun in December 1942, by Defiants carrying "Mandrel" equipment. The Germans countered by altering the Freya wavelength to others which the "Mandrel" could not jam. Other countermeasures introduced were the jamming of the German ground to night fighter communications by transmissions from RAF bombers. Jones, op cit, p.378.

4 Franks, op cit, p.187.

5 Serrate could home on the transmissions of Lichtenstein from 50 miles. The Germans later had a similar device known as the Flensburg homer, which could home on Monica transmissions.

6 Jones, op cit, p.504.

landed in error at RAF Woodbridge in Essex on 13 July 1944, was to reveal more German secrets and enabled valuable information to be gathered[7] Fitted with the *Flensburg* radar detection equipment, the German night fighter was able to home on a bomber's Monica transmissions from a range of fifty miles to within 1000 feet.[8] After Monica's compromise by *Flensburg*, it was withdrawn from use from the main force of Bomber Command during 1944, but continued in use with the night fighters.[9]

Although not fitted with any of the black box wizardry of the 100 Group aircraft, 456 Squadron flew its first Mahmoud patrol on the night of 25 November 1943. Deployed to the forward base of Bradwell Bay in Essex, Pilot Officer "Tommy" May[10] and Flying Officer Les Parnell (HJ650), took off in the middle of the night, only to be recalled due to unsuitable weather. The following evening, Pilot Officers Bob Pahlow[11] and Frank Silva[12] (DD620) crossed the French coast near Dunkirk, and in the face of enemy jamming of their radar, returned to Bradwell Bay. These sorties continued into the next month. On the evening of 2 December, Flying Officers Keith Roediger[13] and Bob Dobson[14] (DZ682) departed for a patrol to Berlin, the deepest penetration of the Continent that had been planned at that stage for 456 Squadron. Crossing over the Hook of Holland at 23,000 feet, Dobson soon had a contact off to the starboard at almost maximum range. Turning and holding the contact for several minutes, the blip then faded - targets were now fast and highly elusive - so they altered back onto course. After almost half an hour, Dobson then obtained another contact at nearly five miles, and they were able to close to 2,000 feet. At this critical stage, the starboard engine began to emit sparks, and as its temperature rose the chase was abandoned. Soon all power was lost from the engine, and it was feathered, forcing Roediger to return from Germany to Essex on one engine.

The following night, 3/4 December, two Mahmoud sorties were launched. Pilot Officers Gate Gatenby and Scotty Fraser (DZ293) had engine trouble half way to the Dutch Coast, and had to return. Pilot Officer Tommy

7 Saunders, op cit, p.265.

8 Jones, op cit, p.586.

9 The Germans developed the Naxos tail warning device from AI radars which was also capable of enabling their night fighters to home for the bombers' H2S navigation radars from ranges of 40 miles. Immediately, Bomber Command instituted an emission control policy, limiting the transmission of their radars to only minimal use for checking navigation and for their bombing runs.

10 Plt Off J L May, 412993. No 456 Sqn. Student of Sydney; b Kurri Kurri, NSW, 14 Nov 1922. Killed in action 4 Dec 1943.

11 Flg Off R M J Pahlow, 411255. No 456 Sqn. Agricultural field officer of Leeton, NSW; b Sydney 1 Feb 1917. Killed in action 29 Apr 1944.

12 Flg Off F M Silva, 413267. No 456 Sqn. Clerk of Sydney; b Liverpool, Eng, 26 Jun 1922. Killed in action 29 Apr 1944.

13 Flt Lt K A Roediger, 415227. No 456 Sqn, TFU Defford. Butcher of Northam, WA; b Balaklava, SA, 25 May 1921.

14 Flt Lt R J H Dobson, 413179. No 456 Sqn, TFU Defford, O'seas HQ. Schoolteacher of Kandos, NSW; b Sydney 24 Aug 1910.

Mosquito DD739, which failed to return from a Mahmoud with Tommy May and Les Parnell.

May and Flying Officer Les Parnell (DD739) continued towards Germany on their patrol to Kassel, south of Hannover, and nothing more was ever heard from them. It was on this sad note that 456 finished its Mahmouds. Mahmoud, from the outset, had only been considered a temporary measure, and was to be superseded as later equipment and aircraft became available.[15] When 456 Squadron did return to bomber support duties in 1945, it was in a slightly different role.

The new CO, Wing Commander Keith Hampshire[16], assumed command of 456 Squadron on 14 December. Hampshire, a colourful character who had commanded No 22 Squadron RAAF flying Bostons in New Guinea, had already been awarded the DSO. He was very competitive, and soon had a reputation of being a hard boss, inaugurating a "thorough" regime, which after a long period of virtual inactivity, seemed burdensome at first. As one of the Squadron navigators, Flight Lieutenant Merv Austin, recorded:

> He conceived his purpose in the simplest terms - to weld the Squadron into an efficient fighting unit to enable it to discharge the responsible task that was to be assigned to it in the spring and summer of 1944. He carried out his purpose by the power of his example and the vindication in practice of his methods of infinite care right down to the most meticulous detail. He made the Squadron realise its organic unity - the independence of air and ground crews - and drew every member into the combined effort by giving him a sense of his function and responsibility, by providing opportunities for pooling ideas, but above all by infusing a new sense of drill and discipline. It is neither excessive nor invidious to

15 Sharp & Bowyer, op cit, p.300.

16 Gp Capt K M Hampshire DSO* DFC, 147. Nos 2 & 12 Sqns, SHQ Darwin; comd 6 Sqn 1941-42, 23 Sqn 1942, 22 Sqn 1942-43, 456 Sqn 1943-44, 280 Wg RAF Transport Cd 1945. Regular air force offr of Perth; b Port Macquarie, NSW, 10 Sep 1914.

say that he contributed more to the making of the Squadron than any other individual.[17]

Hampshire's work was not accomplished without a certain amount of stress and friction. Good leaders aren't always popular, but the Squadron was to achieve its greatest results under the period of his command.

The new CO, Wing Commander Keith Hampshire.

The New Year saw a series of scrambles from Fairwood Common, in abysmal weather, against reported raids which never ventured far enough towards 456's sector. In addition to now having a new adjutant, Flight Lieutenant "Tony" Luxton[18], the Squadron had also received several Royal Navy crews. The Fleet Air Arm was to form night fighter squadrons, so three crews[19] were posted to the RAAF squadron for experience. The only serious accident that occurred was on 15 January, when Flying Officer Ted

17 Flt Lt M N Austin, op cit, Part IV, p.9.

18 Sqn Ldr W H Luxton, 400133/O35926. No 502 Sqn RAF, 461 & 456 Sqns, 21 Sqn. Company director of Melbourne; b Melbourne 2 Jul 1919.

19 Lt D G Thornley/SLt D R H Phillips, Lt A J Y Brown/SLt R R Milton-Hine, and SLt Ogston/SLt J C Harrison. The last crew did not remain long with 456, but the others served from Nov 1943 until Sep 1944.

Radford[20] and Pilot Officer Wally Atkinson[21] (DZ681) were landing with an engine problem, and overshot the runway. Running off the end, and down into a deep valley, the crew was unhurt, but the aircraft was written-off.

The advent of 1944 had also seen an updated model of the Mosquito delivered to 456 Squadron. One hundred NF.II aircraft had been converted to carry the American SCR720 radar, known to the RAF as AI Mark X, and designated Mosquito NF.XVII. The first five were received on 456 Squadron on 29 January 1944. Because of the high security of the new AI set, these aircraft carried serial numbers with a G suffix (eg HK246/G) to indicate the need for armed guarding at all times. In fact, aircraft fitted with Mk X AI were considered so sensitive, with the fear that the magnetron valve (the core of centimetric radar) might fall into enemy hands, that they were forbidden to fly over enemy territory until 5 June 1944, the eve of D-Day. The fitting of the AI Mark X in the bulbous "universal" radome had meant that the four nose machine guns were not carried, relying only on the firepower of the four cannon.

The re-equipment with centimetric radar required the visit to Fairwood Common of the AI "circus", a flight equipped with Wellington bombers converted to AI radar trainers, to train the nav/rads in the operation and techniques of the new equipment. The AI Mark X, as described in the previous chapter, only swept in the forward sector, so the Mosquito XVIIs were also equipped with Monica tail-warning radar equipment to warn of the approach of an enemy fighter from behind. Flying one of the Wellingtons as part of the "circus", was an old 456 hand on his rest from operations, Pilot Officer Johnny Newell.

Flying Officer Fred Stevens had returned to 456 during January to commence his second tour of operations. His observer, Mark Stokes, had recently been killed in a flying accident during his "rest" tour at 54 OTU, so Stevens teamed up with a new navigator, Pilot Officer Andy Kellett[22]. Also returning for another tour with 456 was the new 'B' Flight Commander, Squadron Leader Bas Howard, who was crewed again with Flying Officer Jack Ross. Their arrival coincided with the news that the 'A' Flight Commander, Squadron Leader Geoff Robinson had been awarded the DFC. His award cited his previous operational flying with the RAF, and the leadership qualities he had displayed while serving on 456 Squadron.[23]

Uneventful scrambles were carried out during February, with the first operations being flown by the new aircraft on 12 February. There was now an increased state of readiness, with four aircraft being required for each of the Colerne and Fairwood Common sectors. In addition, as a handful of the

20 Flg Off E C Radford, 415549. No 456 Sqn. Pastry cook of Mt Lawley, WA; b Perth 26 Oct 1922. Killed in action 12 Jul 1944.

21 Plt Off W E Atkinson, 422530. No 456 Sqn. Clerk of Sydney; b Bathurst, NSW, 27 Aug 1919. Killed in action 12 Jul 1944.

22 Flt Lt W A H Kellett, 410343. No 456 Sqn, RAAF HQ. Clerk of Melbourne; b Melbourne 4 Sep 1920.

23 London Gazette No 36392, 22 Feb 1944.

older Mosquito IIs had been kept on the Squadron, several Ranger sorties were undertaken. On the night of 10 February, a pair of aircraft staged through Predannack for sorties to Bordeaux. Wing Commander Hampshire and Flying Officer Tom Condon (DZ293), and Flying Officers George Houston and Lars Engberg (DD622), flew separate routes on these five-hour Rangers, both successfully finding their targets and attacking aircraft hangars at Lake Hourtin and Lake Biscarrosse.

The following night, 11 February, Flight Lieutenant Bob Lewis[24] and Pilot Officer Bob Wilmoth[25] (DD622) were forced back from France in poor weather, and weather on the night of the 17th also curtailed a repeat performance of Biscarrosse Ranger. A week later, Flying Officers Tony Samson and Alec Abbey (DZ293) staged through Exeter for an uneventful Ranger to Rennes airfield, but on return the next night, their luck changed. Coasting into France at Mont St Michel, they flew low to Rennes, arriving at 1900 hours. After a wait of ten minutes, they spotted a Ju 88 approaching the airfield from the north-west, with its wheels and flaps down. Closing in to 100 yards, with the enemy aircraft at only 30 feet off the ground, Samson fired and saw strikes along the wings, between the engines and on the fuselage, setting the aircraft alight. As they left the airfield, they could see the wreckage burning on the ground.

Such successes gained by the RAF night fighting Mosquitos attacking the enemy at his airfields had an unnerving effect on *Luftwaffe* crews. They knew that at any stage of their sortie, from take-off to landing, they were prey to the far-ranging Mosquitos. The result was that the two-crew Bf 110 night fighters were now carrying a crew of three - pilot, radar operator and, now, rear gunner.[26]

24 Flt Lt W R V Lewis, 407969. No 488 Sqn RNZAF, 456 Sqn, 51 OTU. Bank clerk of Adelaide; b Adelaide 12 Jun 1922.

25 Flt Lt R S Wilmoth, 401594. No 488 Sqn RNZAF, 456 Sqn, 51 OTU. Draughtsman of Melbourne; b Melbourne 13 Nov 1913.

26 Bekker, op cit, p.335.

CHAPTER ELEVEN

THE GERMAN PERSPECTIVE

The Germans had made similar strides to the British in development of a defensive radar system. Their early warning *Freya* radar network had been deployed extensively during 1940 and 1941. Unlike the British Chain Home system, the *Freya* was mobile, and from positions along the coast provided plots on air and sea targets. *Freya*, the goddess Venus of Norse mythology, could report direction and distance out to nearly 200 kilometres, but could not give altitude.[1] Initially with a pulse repetition rate of 1000 pulses per second (pps), this was subsequently lowered to 500 pulses, which enabled a detection range of 300 kilometres.

In addition to the early warning radar, the Germans had developed a GCI system of smaller *Würzburg* parabolic reflector ground radars, which originally had controlled flak and searchlights. The first *Würzburg*, with a range of 40 kilometres (22 miles), was installed during October 1940, near the Dutch town of Zutphen, north-east of Arnhem. *Würzburg* could track with such accuracy, while also providing the altitude of a target, that in May 1941 a British bomber flying in dense cloud over Essen-Frintrop airfield was downed by radar-directed flak.[2] The system was to form an air defence network adapted to control night fighters. Until this stage, German fighters

1 Freya operated on a wavelength of 240 cm, ie. a frequency of 120 MHz. Bekker, op cit, p.206, and Jones, op cit, p.252.

2 Bekker, op cit, p.145; Bartz, op cit, p.54.

had not operated under ground control, as the nature of their war had been offensive, to the neglect of defence.

By 1941, *Würzburg* radars were co-located with the longer-range *Freyas*,[3] and the standard air defence site was equipped with a *Freya* and two *Würzburg* radars. The *Freya* would first pick up an incoming attacker, and direct one of the *Würzburgs* on to it. The other *Würzburg* would be tracking the scrambled German night fighter, which was orbiting a beacon in front of the defences. Ground control would direct the fighter onto an interception course, while the *Würzburg* tracking the bomber would direct three searchlights towards the intruder. Once this was picked up, a battalion of searchlights in the defensive belt would attempt to visually track the bomber to keep it illuminated for the fighter.

When General Kammhuber was promoted to command all *Luftwaffe* night defences, he planned for two different roles for his night fighters - defensive, in a restricted zone on the German western frontier, and offensive long-range "intruder" operations against the RAF bombers' home bases.[4] However, intruder operations against British airfields had ceased in 1941, as the German leadership assessed the propaganda value of far-away victories over England did nothing to improve the morale at home.

In October 1940, Kammhuber set up three night fighter zones in the Zuider Zee and Rhine estuary, across the path of RAF bombers attacking the Ruhr. The three zones were contained in an area 90 kilometres long by 20 kilometres deep, each occupied by a searchlight battalion and two *Würzburg* radars.[5] This was then developed by summer 1941, when the searchlight belt was supplemented by a string of *"Himmelbett"* zones,[6] one overlapping the next, each with a radius of some 50 miles. A *Himmelbett* station comprised:

> one *Freya* set, with a range of up to 100 miles for early warning,
> one *Würzburg*, for plotting the bomber,
> one *Würzburg* for guiding the fighter.
> one *Seeburg* evaluation table *(Seeburgtisch)*, plotting progress of the two aircraft.

The *Kammhuber Line*, named by the British after the *Luftwaffe* "General of Night Fighters", was a belt of these radar-operated *Himmelbett* ground-control zones, designed to provide protection of the Ruhr industries, which stretched from northern Germany to Liege.[7] By March 1941, the

3 Richards, op cit, p.396.

4 Bekker, op cit, p.207.

5 *The Rise and Fall of the German Air Force,* Air Ministry Pamphlet No 248, Air Ministry, London, 1948, p.186.

6 The name Himmelbett is literally a four-poster bed, referring to the four components of each air defence sector: the Freya radar, the two Würzburg radars, and the Seeburg plotting table.

7 Bartz, op cit, p.109.

Würzburg operated with a small 10-foot diameter paraboloid antenna initially on a decimetric wavelength of 53cm (560 MHz), its high pulse repetition frequency (PRF) of 3750pps giving the system its short range.

Kammhuber Line had reached the Danish border and by the end of the following year had been extended to the south-east of Paris. Kammhuber's ultimate objective was to stretch this line of defence from southern Norway to the Mediterranean.

This string of *Himmelbett* zones, each with a radius adjusted to the range of the *Würzburg,* had one big weakness. Only one night fighter could be operated at one time in one particular zone, or box. This limitation was due to the nature of the German equipment restricting their procedures, by being able to track only one bomber with one fighter in each box. The fighter would orbit suitably placed radio beacons (known as *Kleine Schraube,* "little screw"), awaiting a bomber to enter the box. On the *Seeburg* table the position of the bomber and the fighter were shown as spots of coloured light, red for the bomber, and blue for the fighter, projected from underneath on to the ground glass screen forming the table top. The two projectors were moved by operators receiving instructions from the two giant *Würzburgs.* There were no cathode ray presentations such as the RAF system of GCI, and therefore the plotting system was ponderous and liable to human error.[8] A system of continued pursuit beyond the confining boundariesä of these narrow zones did not exist, and the system could not effectively counter the thousand-bomber raids which began at the end of May 1942.

British jamming of *Freya* early warning radars started in December 1942, by RAF use of the Mandrel jammer. These were carried by Defiants, now relegated to these support tasks, to form an airborne screen for the advancing bombers. With the *Freyas* blinded, the task of the *Würzburgs* was made that much more difficult without knowing the direction of the attackers. The Germans soon countered by changing the frequency of the *Freya*. In addition, 1942 saw the introduction of the Giant *Würzburg*, the *Würzburg-Riese,* which was more accurate, and its large 23 foot reflector aerial gave the system a range of 45 miles (60 kilometres).[9]

Although the *Himmelbett* system allowed for the vectoring of a night fighter to a distance of 400 yards from its quarry,[10] quite often the ground controller brought the two dots together on his plot without the pilot being able to make visual contact. Clearly something was needed to bridge this gap. During the summer of 1940, the Germans had trialled an infra-red device on night fighters to aid detection of targets, but this had not proved successful.[11] They, like the British, then turned to AI radar.

The German electronics company *Telefunken* had developed an AI, *Lichtenstein,* which by July 1941 demonstrated a minimum range of 200 yards, and a maximum range of 3,000 yards. By July 1942, the strength of German night fighters stood at 250 aircraft, organised into *Nacht Jagd Geschwaden* (abbreviated to *NJG*, night fighter wing), armed typically with Messerschmitt Bf 110 and Junkers Ju 88 fighters. The first unit to equip with *Lichtenstein B/C* AI (known to the Germans as *FuG 202*[12]) was the *Luftwaffe's* first night fighter wing, NJG1. The system operated on a frequency of 490 Mhz (62cm wavelength), and searched 25 degree either side of the aircraft nose out to a maximum range of 2 miles.[13] *Lichtenstein C-1 (FuG 212)*, the first major production version, the equivalent of the British AI Mark IV, had a clumsy external antenna array, which reduced aircraft speed by as much as 25 mph.

The RAF then developed a radar warning receiver, the "Boozer", which warned Bomber Command crews when their aircraft were being illuminated by *Lichtenstein* or *Würzburg* transmissions. This came into

8 Jones, op cit, p.345.

9 The PRF was dropped to around 2000pps. Bekker, op cit, p.208; Jones, op cit, pp.294, 360.

10 Air Ministry Pamphlet No 248, p.191.

11 Air Ministry Pamphlet No 248, p.188.

12 Abbreviation of Funkgerät, radio equipment.

13 Bartz, op cit, p.108; Howard-Williams, op cit, p.177.

service in the spring of 1943, by which time the *Luftwaffe* night fighter force had grown to a strength of 490 twin-engined aircraft, equipping five *Geschwaden*.[14]

The counter to *Lichtenstein* and *Würzburg* then came with the RAF attempting to jam German air defences by the use of "Window" on the night of 24 July 1943. Window was tin foil strips, developed by the British during 1942, but had not been initially used for fear of retaliation before an antidote had been devised.[15] The Window foil, cut to exactly half the wavelength of the *Würzburg*,[16] was made up in packets each weighing a pound, and dropped out of the leading aircraft of a formation at the rate of one pack per minute to form a blinding screen. This caused so much interference that the radars were paralysed. When Window was dropped, real targets could still be identified on *Freya*, but the *Würzburg* radars became indecipherable and reduced the system to chaos, as control of the night fighters was dependent on position and height information, and without it, the controllers were powerless. With the RAF bomber streams protected by Window, the *Himmelbett* system became useless overnight, and the night fighters could only fumble in the dark. To counter the jamming, *Würzburg* radars operated on different wavelengths up to 67cm.

A new German AI was already in development when *Lichtenstein B/C* was rendered useless by Window. From October 1943, *Luftwaffe* night fighters acquired *Lichtenstein SN2* (the *FuG 220*), which operated on a wavelength of 330cm (a frequency of 85 MHz), therefore requiring an even larger cumbersome aerial array, resembling antlers. Its advantage was that *Lichtenstein SN2* was impervious to the existing Window, being able to see through the foil strips.[17] *SN2* also had the improvement of a much wider angle of search (120 degrees).

Lichtenstein SN2 had two tubes, the left indicating direction of flight, the right the target height. The aim was for the radar operator to get the fighter in a position behind the target, so the pilot could then obtain a visual acquisition by silhouette or from the exhaust flames. However, although the *SN2* had an improved maximum range out to 6 kilometres (4 miles), it did not function within 400-500 metres (1300 to 1650 feet).[18] This was inadequate to ensure visual identification, and soon led to this more modern system being supplemented by a simplified version of the old *FuG 202* with a single quad dipole aerial, improving the problem of minimum range and allowing closure to around 300 feet.[19]

---

14 Air Ministry Pamphlet No 248, p.191.

15 Jones, op cit, p.375, and D Richards & H StG Saunders, *Royal Air Force 1939-1945*, Vol *II*, HMSO, London, 1954, p.157.

16 Window was cut in lengths of about 10" (ie 26 cm, half wavelength), and 3/4" (1 to 2 cm) wide.

17 Bekker, op cit, p.335; Bartz, op cit, p.117; Gunston, op cit, p.103.

18 Howard-Williams, op cit, p.178; Bekker, op cit, p.337; Gunston, op cit, p.103.

19 Gunston, op cit, p.104.

German night fighters were also directed by a ground "running commentary" system. This procedure was supplied by the German Observer Corps, who provided details of the plot of the bomber stream, and the transmission of this information provided the running broadcast. The commentary provided the height and direction of the bomber stream, and the area of which it was passing, which acted as the eyes for any night fighters within range. The German night fighters adopted this *Zahne Sau*, Tame Boar, technique which placed their night fighters in the bomber stream as early as possible, and kept them there for as long as possible.[20] Another development of these "freelance fighters", as "Bomber" Harris called them, was the use of searchlights, simultaneously in large numbers, for the illumination of the cloud base, and the later use of flares, with night fighters operating above the flak and searchlight defences.[21] This was known as the *Wilde Sau*, or Wild Boar, tactics.

Other electronic wizardry developed by the Germans included *Naxos, Korfu* and *Flensburg. Naxos-Z (FuG 350)*, which came into operational *Luftwaffe* service in January 1944, enabled German pilots to home onto the H2S radar transmissions of the bomber stream from a range of 50 kilometres (30 miles), but was not precise enough for individual target selection. Furthermore, the Germans also perfected *Korfu*, which enabled them to detect H2S transmissions from the ground. A *Korfu* network was spread throughout the entire air defence system of western Germany. H2S transmissions coming from bombers parked on their bases in Britain could be detected in Germany, and the position of any H2S bomber stream could constantly be plotted from the time of its take-off until its landing.[22]

Flensburg *(FuG 227)*, carried by most German night fighters by the end of 1943, was a passive receiver which homed on the transmissions of the RAF bombers' Monica tail-warning radar. On 13 July 1944, the first *SN2* was captured by the Allies when a Ju 88G night fighter, of NJG2, landed in error at Woodbridge, in Essex. When this aircraft was evaluated by the RAF, it was found that *Flensburg* gave clear bearings onto a radiating Monica from an incredible distance of 130 miles (200 kilometres).[23] Bomber Command was then to issue the order for its aircraft to leave Monicas switched off. It was in the face of this type of technology that led to the RAF forming No 100 Group for bomber support missions in November 1943.

In spite of the strides made by the German night fighter force, possibly its greatest failing was not taking the battle to Bomber Command's air bases. As early as July 1940, the night fighter force had commenced night intruding operations against RAF Bomber Command airfields in Yorkshire, Lincolnshire, East Anglia and the Home Counties, in addition to attacking

20 Streetly, op cit, p.219.

21 H.StG. Saunders, *Royal Air Force 1939-1945, Vol III*, HMSO, London, 1954, p.11.

22 *The Strategic Air Offensive Against Germany 1939-1945*, Vol IV, Sir Charles Webster & N Frankland, HMSO, London, 1961, p.13

23 Gunston, op cit, p.125.

Lichtenstein B/C AI aerials on a Ju 88

A later model Ju 88 with Lichtenstein SN2

bombers returning from raids across the Channel. In the autumn of 1941, this intruder force was withdrawn to the Mediterranean, and intruder operations over England ceased. Meanwhile, the RAF had adopted a complicated take-off and landing system that would have been highly vulnerable to intruder operations. The *Luftwaffe's* failure to exploit this opportunity, enabling RAF bases and night training to continue operating without interference, left British airfields undisturbed from late 1941 until early 1945, when intruder operations were undertaken for a brief period of only two nights. This lack of a continued night offensive against RAF bases was an important factor which contributed to the final crippling of Germany.

The combination of the SN2 and the smaller Lichtenstein B/C on a Bf 110G night fighter

CHAPTER TWELVE

PREPARING FOR INVASION

On 15 November 1943, Fighter Command had been split into two different formations, with two distinctly different missions. Most of the Command were dedicated to the RAF's 2nd Tactical Air Force, which with the US 9th Air Force, provided the tactical air assets of the Allied Expeditionary Air Force for the invasion of Europe. The remnants of Fighter Command, required for home defence, became known as Air Defence of Great Britain (ADGB). It was not considered at this stage the *Luftwaffe* could launch any serious offensive against southern England, and the most that could be attempted was thought to be between 100 or 150 sorties by German bombers in any one night.[1]

During February 1944, that was exactly what the *Luftwaffe* had done, when it launched "Operation Steinbock", a mini Blitz of London and the south. No 456 Squadron, situated to the west in Wales, had been out of the action, so at the end of the month, the Squadron was directed to Ford, in Sussex, to join 11 Group in its efforts in defeating the offensive. The CO, Keith Hampshire, led the formation of 16 Mosquitos from Fairwood Common to 456 Squadron's new base at Ford on Tuesday 29 February.

Ford Junction, midway between Arundel and Littlehampton on the Sussex coast, had been built as a First World War training station for the Royal Flying Corps, and having been a Royal Navy airfield at the beginning of the Second World War, was taken over by 11 Group in October 1940. All-weather operations had been helped by construction of two tarmac

1 Saunders, op cit, Pt III, p.98.

runways of 6,000 feet (SW/NE) and 4,800 feet (NW/SE), taxi-tracks, with blast pens and open-ended "blister" hangars around the perimeter track. Having replaced No 29 Squadron, 456 now shared the base with another Mosquito squadron, No 418, a Canadian unit engaged in Ranger operations. The Officers' Mess was situated about a kilometre north of the base, in the Tortington Girls' School. Operations were controlled by the nearby GCI at Durrington, often in conjunction with searchlight instructions from the coastal searchlight belt. This procedure was known as "gauntleting", whereby the fighter would hold over a beacon or a flashing morse-coded light, awaiting searchlight directions on where to head to conduct an AI search, followed by illumination of a target by the searchlights.

The Squadron was into action immediately. On the night of 1 March, 50 to 60 enemy aircraft raided Sussex and Kent, with about a dozen reaching London. Six aircraft were on airborne patrol, when Pilot Officers Bob Richardson and Tom Landy (HK282) were gauntleted onto an illuminated target that they identified as a Dornier Do 217 bomber. Richardson closed and fired a burst which struck the fuselage and port wing root, but with his high closing speed, he then overshot the target. Flying Officers Col Griffin[2] and Harry "Hoppy" Williams[3] (HK312), flying at 22,000 feet, obtained an AI contact and they closed stealthily to 700 feet. The sheet of cloud below at 10,000 feet was well illuminated by the searchlights, a situation requiring night fighters to be wary. Griffin visually identified the target as a Do 217 by its twin fins, just as it jinked into a hard turn to the right. Losing visual contact immediately, they were unable to find it again on AI.

Occasional enemy raids ventured over Sussex through the rest of the month, but there were no more engagements until the night of 21 March. One of the Fleet Air Arm crews, Lieutenant Dennis Thornley and Sub Lieutenant Derrick Phillips (HK359), being controlled off French coast by Black Gang GCI on the Isle of Wight, obtained an AI contact, which they closed and visually identified as a single-engined Focke Wulf Fw 190 fighter. Thornley saw his bullets strike home on the port side of the enemy fighter's engine, sending back sparks through the Mosquito's propeller. The fighter dived steeply towards the sea, with the Mosquito firing at it again, but as no crash was witnessed, the crew could only claim a "probable".

Meanwhile, Flying Officers Keith Roediger and Bob Dobson (HK297) had been gauntleted onto a target approaching the London inner defences, and had to break off the chase when the anti-aircraft belt opened fire. They then turned to the south, where there was an intersection of about 20 searchlight beams, and Dobson was able to get an AI contact at 7 miles. On closing, the target took violent evasive action, climbing and diving 5,000 feet at a time. In a dive, the Mosquito crew was able to close to 150 yards and clearly saw the target's German markings, and identified it as a Ju 88. Roediger fired a burst, causing the bomber to peel off, but the searchlights

2 Flt Lt C S Griffin, 416568. No 456 Sqn, 51 OTU, 456 Sqn. Police instructor of Adelaide; b Strathalbyn, SA, 1 Sep 1919.

3 Flt Lt H P Williams, 26403/285145. No 456 Sqn, 51 OTU, 456 Sqn, CFE. Bank clerk of Adelaide; b Adelaide 2 May 1920.

Aircrew at Ford, February 1944
From left: Dick Wilkie, 'Gate' Gatenby, John Harrison RNVR (standing), Ray Milton-Hine RNVR (sitting), Bob Wilmoth, John Stickley, Bob Dobson, Stan Williams, T C Kewen RCAF, Ted Radford.

soon picked it up enabling the Mosquito to close again to 300 yards. Evading the searchlights again, the bomber was then held on AI, and was fired on again. Hits were seen on the port wing root, followed by a large flash. The bomber then dived steeply from 16,000 feet, pursued by the Mosquito, but was lost in cloud at 5,000 feet. Shortly afterwards, the searchlight site at Rye, in East Sussex, saw a diving object on fire hit the sea.

Three nights later, on 24 March, about 120 German raiders crossed the coast near Ford and the Mosquitos were scrambled. Soon the sound of cannon fire could be heard by the groundcrew and strikes could be seen on a stricken enemy bomber. It was a Ju 88 bomber, from *Kampfgeschwader* 6, which the CO and Tom Condon (HK286) brought down at Walberton, just to the north of Ford aerodrome near Arundel. The Australian crew had closed on their AI response, and with the aid of binoculars (fighter crews had now been supplied with Ross night binoculars), had identified the target as a Ju 88. Hampshire then slid his Mosquito to the port side to avoid debris, and gave a short burst from 150 yards. The enemy aircraft disintegrated in flames, with the port wing and engine falling off, and one of the four crew was able to bale out safely to be taken prisoner. Flying Officer Fred Stevens and Pilot Officer Andy Kellett (HK359) had chased an Me 410 and were fired at, but could not get within range of the violently evading German fighter, which they pursued to the French coast.

Andy Kellett and Fred Stevens

On the night of 27 March, Bristol was the target for 100 enemy bombers, enabling No 456 Squadron to achieve its greatest success to date. The CO, Wing Commander Hampshire, with Flying Officer Tom Condon (HK286), scored two Ju 88s, which both crashed on the land near Portland. Their first, again belonging to KG6, burst into flames and again one crewman was taken prisoner. The second aircraft, which Hampshire attacked at midnight, was a Ju 88A-4, coded B3+BL, belonging to 3 *Staffel* of KG54. Despite being despatched in flames by the Mosquito, three crewmen were able to safely bale out of the disintegrating wreckage. The parachute of the fourth crewmember, the wireless operator, had failed to open. The *Luftwaffe* crew were:

Pilot - *Oberfeldwebel* Hans Bräutigam
Observer - *Obergefreiter* Kurt Chalon
Air Gunner - *Obergefreiter* Alfred Maletzki
Wireless Op - *Unteroffizier* Robert Belz.

The bomber, armed with 2,000kg of incendiary bombs, had flown from its French base at Juvincourt, near Reims, across the north of Paris to Cherbourg, and then attacked the docks at Bristol, where German Intelligence expected there to be a buildup of material in preparation for the Allied landings. Kurt Chalon, the observer, later recalled:

Crossing the English coast we met some searchlights. Only some A/A guns were firing. No nightfighters seen. Then the attack of the night-

> fighter came so suddenly that we had no chance for any defence. The nightfighter was directed by radar. We could hear that moments before the attack [most probably a reference to radar warning equipment]...The same moment the nightfighter attacked our plane we could - pilot and me - see the silhouette of it, so we knew that it was a Mosquito. The gunfire of all 4 guns of the nightfighter hit our left engine, it was burning immediately. The pilot gave the order to get out at once.[4]

The surviving crewmembers joined up on the ground, and evaded for the night, catching several hours of sleep in a small hut. Setting off in the morning at 0600 hours, they were apprehended by the police two hours later. The air gunner, Alfred Maletzki, later wrote:

> On my capture I handed my belt with the pistol to the policeman and told them: "I am a German flyer, who just was shot down the last night". They offered me cigarettes, since I don't smoke I ask them for something to eat, because I was quite hungry. The box of pills which should keep the hunger away did not work well with me. At Bridgewater I got my first meal...they searched me, took my belt and told me to keep my zipper down on my flying boots. For me this was a laugh, since I had no intention for escape. In Taunton I was told of the fate of Robert Belz.[5]

As the German aircraft had blown up, it had sent debris back onto the Mosquito, damaging its starboard wing. The CO found his aircraft was now uncontrollable in starboard turns, but he still pursued another violently evading target before having to give up the chase. A low speed handling check revealed that his aircraft stalled at 160mph, but he skilfully returned for a safe landing. Meanwhile, the 'B' Flight Commander, Squadron Leader Bas Howard, with Flying Officer Jack Ross (HK323), had been vectored by Durrington GCI to investigate a low-flying bogey heading towards Cherbourg. Howard closed to identify a Ju 88, which he sent down in flames into the sea.

March had been successful, with five confirmed victories, a probable and a damaged, and trade continued for 456 during April. During the month Squadron Leader Geoff Howitt DFC[6] took over 'A' Flight from Robbie Robinson, who left to command No 52 OTU. Howitt brought with him his navigator, Flight Lieutenant George "Red" Irving DFC[7].

18 April 1944 was the last major raid by the *Luftwaffe* on London, but

4 Kurt Chalon, letter, 22 Feb 1972, RAAF Personnel Records, file of Gp Capt Hampshire.

5 Alfred Maletzki, letter, 6 Aug 1972, RAAF Personnel Records, file of Gp Capt Hampshire.

6 Wg Cdr G L Howitt DFC* RAF, 81037. Nos 245, 615 & 85 Sqns RAF, 51 OTU, 85 Sqn RAF, 63 OTU, 456 Sqn; comd 125 Sqn RAF 1944-45. Regular air force offr of Brighton, Eng; b Wallington, Surrey, 29 Jan 1914.

7 Flt Lt G N Irving DFC RAF, 121046. No 85 Sqn RAF, 63 OTU, 456 Sqn, 125 Sqn RAF. Public servant of Carlisle, Cumberland, Eng; b Carlisle 29 Oct 1919.

The CO, left, and Flt Lt Tom Condon examining the wreckage of the Ju 88A, 3E+AP, they brought down the night of 24 March near Ford.

attacks on the south coast were maintained. It was from here, of course, that the Germans anticipated any invasion would be launched. The following night, twelve enemy aircraft were operating over Kent, Surrey and Sussex, when Flight Lieutenant Clive Brooks[8] and Warrant Officer "Freddie" Forbes[9] (HK253) were vectored by Durrington GCI onto a bogey approaching the coast at 24,000 feet. Turning in behind with too much overtake, Brooks obtained a visual on a Me 410, and managed to slow enough to get into a firing position. Firing a 4-second burst from 150 yards, the Me 410's starboard engine and fuel tanks caught fire, and the aircraft rolled onto its back and went straight down in flames, crashing south of Horsham, in Sussex.

About 40 enemy aircraft operated over the south coast on the night of 23 April, when Wing Commander Hampshire, with Tom Condon (HK286), controlled by Black Gang GCI, had success once again. Condon gained an AI contact to the left, at 3 miles, and Hampshire closed to gain a visual at 800 feet, identifying a Ju 88 bomber. Now flying near Swanage, in Dorset, Hampshire sent several bursts into the weaving bomber, which fell away to the left into a steep dive, then burst into flames with pieces of glowing debris falling off. They then chased another Ju 88, but a momentary failure of the cannon prevented a further kill. Pilot Officer Ivor Sanderson[10] and Sergeant Charles Nicholas[11] attacked another Ju 88, which peeled off immediately and evaded further fire.

Two nights later, 25 April, a force of 60 German aircraft approached the coast near Beachy Head, while others operated on mine-laying over the Channel. Flight Lieutenant Lewis and Flying Officer Wilmoth (HK317) destroyed a Ju 188,[12] which exploded, near Portsmouth. Flying Officers Roediger and Dobson (HK297) were scrambled onto a second wave of raiders, and a visual was obtained on another Ju 188. Firing from 150 yards, the bomber's starboard engine exploded, and the aircraft dived in flames, crashing in the sea off Portsmouth. Flying Officers Houston and Engberg (HK264) sent a Ju 88 down in flames nearby.

The CO, flying this sortie with Flying Officer Stew Smith (HK286), was scrambled the night of 28 April, to intercept low-flying targets, suspected of minelaying in the Channel. Positioning up moon, they soon had a contact, and descended to 500 feet over the water. Hampshire was soon able to recognise the shape of a Do 217, and fired a burst into the wing root and fuselage. The German turret gunner returned fire, missing the Mosquito. After a second burst from the Mosquito, the German spiralled away to the

8 Wg Cdr C L Brooks RAF, 47998. 51 OTU, 456 Sqn, RAE Farnborough. Regular air force offr of Fakenham, Norfolk, Eng; b Holt, Norfolk, 22 Oct 1920.

9 WOff R J Forbes RAF, 817174. 51 OTU, 456 Sqn.

10 Flg Off I W Sanderson DFC, 413670. No 456 Sqn, HQ 10 Gp. Farmer of Condoblin, NSW; b Trundle, NSW, 3 Jun 1920.

11 Plt Off C H Nicholas RAF, 1622548/184454. No 456 Sqn.

12 September 1943 had seen the introduction of the improved Ju 188 bomber, a faster and more powerful version of the Ju 88, which was capable of carrying double the load over almost twice the distance. Saunders, op cit, p.409.

Tom Condon, who flew with Wg Cdr Hampshire while he commanded 456 Squadron

left, which Hampshire was unable to safely follow. No crash was seen, so the CO had to be satisfied with a probable.

However, this night's activities had not been without cost. Flying Officers Bob Pahlow and Frank Silva (HK321) had been airborne at the same time as the CO, and had chased a low-flying target off the coast. Black Gang heard no more of their transmissions after 0034 hours, and wreckage of the aircraft was later found in the sea. Nevertheless, April had been another successful month, with a score again of five confirmed, with a probable. During both March and April, 456 was amongst the top scoring night fighter units in home defence.

After a two week break from raids, and under a "bomber's moon" on the night of 14 May, 100 *Luftwaffe* aircraft mounted a coordinated attack on Bristol, Portsmouth and Southampton. Flying Officers Arthur McEvoy and Merv Austin (HK246) were put onto a contact west of the Isle of Wight, which they eventually identified as a Ju 188. Their 2-second burst was sufficient to blow up the whole starboard side of the bomber, which dived in flames from 18,000 feet and crashed at Larkhill, near Netheravon on the Salisbury plain, in Wiltshire. The next night, the *Luftwaffe* attacked Portsmouth, where Flying Officers Doug Arnold[13] and John Stickley[14] (HK297) found a Ju 88. They scored hits on the bomber's port side, followed by an explosion, sending the doomed aircraft down to crash near Medstead, in Hampshire.

13 Flt Lt D W Arnold, 418045. No 456 Sqn. Clerk of Melbourne; b Melbourne 28 Jul 1921.

14 Flt Lt J B Stickley, 425374. No 456 Sqn. Commercial traveller of Rockhampton, Qld; b Rockhampton, 14 Jun 1915.

Another raid on Portsmouth on 22 May provided more trade, with the CO and Flying Officer Condon (HK286), and Pilot Officer Sanderson with Sergeant Nicholas (HK353), each destroying a Ju 88 off Southampton. It had been another excellent month for the Australian night fighters. With the enemy concentrating on the south coast, Ford was the place to be, in the thick of the action. Now joining 456 Squadron were Flight Lieutenant Bob Cowper DFC[15] and his navigator Flying Officer Bill Watson[16], who had both already completed a successful night fighter tour in the Middle East.

During these busy nocturnal activities of April and May, No 456 Squadron was also witnessing a huge aerial buildup at their aerodrome. Ford was being prepared for use by the Allied Expeditionary Air Force, and complete wings of tactical fighters moved onto the airfield. There was insufficient accommodation for such a force, so the personnel for this airborne armada were living under canvas in preparation for the invasion. During the tactical deployment to the Continent, these fighter squadrons, which included No 453 Squadron RAAF, were to live a nomadic lifestyle. Shortly both these RAAF units, 453 and 456 Squadrons, would be involved in the greatest invasion in history.

On 12 May 1944, Wing Commander Hampshire's DFC was gazetted:

> Wing Commander Hampshire has commanded the squadron since December, 1943, and has displayed inspiring leadership, great skill and gallantry.
> Since operating in this country, Wing Commander Hampshire has shot down three enemy aircraft at night, two of them in one sortie. In the latter of his two fights his aircraft was damaged by flying portions of disintegrating enemy aircraft. Nevertheless, Wing Commander Hampshire flew safely to base, although at times his aircraft was almost uncontrollable.
> His achievement was worthy of the greatest praise and was typical of the determination he has shown throughout his operational career.[17]

15 Sqn Ldr R B Cowper DFC*, 407666. Nos 153, 89 & 108 Sqns RAF, 63 OTU, 456 Sqn; comd 456 Sqn 1945. Draughtsman of Adelaide; b Broken Hill, NSW, 24 Jun 1922.

16 Flt Lt W Watson DFC RAF, 49287. Nos 153, 89 & 108 Sqns RAF, 63 OTU, 456 Sqn. Regular air force offr of Glasgow.

17 London Gazette No 36511, 12 May 1944.

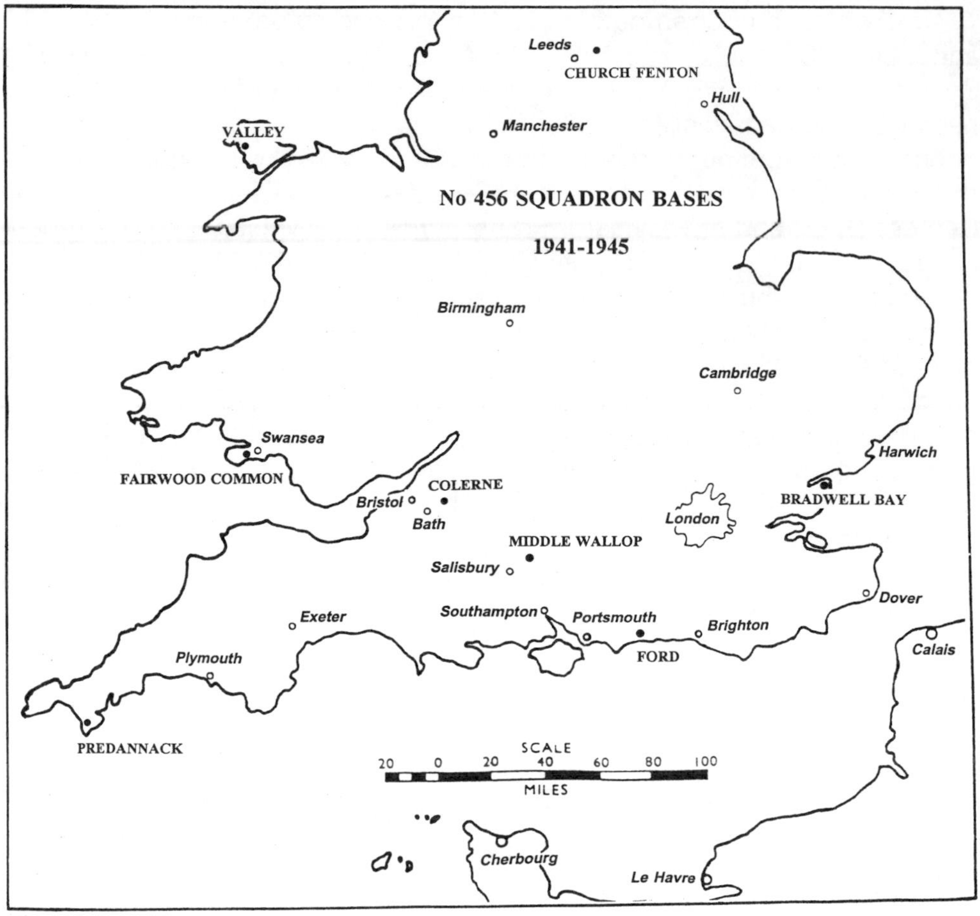

456 Squadron bases 1941-1945

CHAPTER 13

D-DAY

With the approach of "Operation Overlord", the imminent invasion of Europe, USAAF night fighters began to arrive at East Anglian and south coast bases. The specially designed US aircraft for night fighting was the Northrop P-61 Black Widow. One unit moved to Hurn, to the west of Ford, and soon after its arrival, the American personnel lined up on the perimeter track saluting a Black Widow taxying out. One commented: "That's the Major, he's going to get the DFC for this; it's the first operational night fighter sortie in the European Theater of Operations."[1]

Well, it was not exactly the first of the war, as the RAF had by now been conducting night fighter sorties for several years. But for the Americans, it was *their* first, and they were now in the UK in force for D-Day operations, the greatest sea and airborne invasion ever attempted. Fighter cover for the Overlord assault would amount to 171 fighter and fighter-bomber squadrons, which was more than sufficient to deal with any German attempt to interfere with the invasion from the air.[2] In fact, the battle in the air had been won before that on the ground had begun.

The mobility that would be required for the supporting fighter units dictated that the Tactical Air Force squadrons be split into two distinct components: aircrew and ground staff. Thus there was 453 Squadron and 6453 Servicing Echelon, under the same CO. The aircrew could move around, landing at different airfields or advanced landing grounds, without the need to move their groundcrew. Maintenance would be performed by the local servicing echelon. No 456 Squadron had already been working with a servicing echelon, No 3064, which had provided major servicing as

1 Meadows, Pt 2, p.20.

2 Saunders, op cit, Part III, p.97.

The Heinkel He 177A - the "flying firework". This one had been captured, and flown with an RAF serial number, TS439, on evaluation.

part of the Squadron since the early days in Valley. In line with the tactical units, on 22 March 1944 this was renumbered 6456 Servicing Element. Leading the servicing echelon was the Squadron's energetic engineering officer, Flight Lieutenant Stu Green[3], who would ultimately not only be mentioned in dispatches for his efforts, but also awarded the OBE.

One aircraft that the aircrew were briefed they would encounter during the invasion was the Heinkel He 177 four-engined bomber, which carried the Hs 293 radio-controlled glider bomb.[4] These flying bombs, the first air to surface precision guided munitions, were to be unleashed onto the invasion fleet. The He 177 had a unique engine arrangement, with two engines each side being coupled to drive a single propeller. The bomber carried a crew of five, had a speed of over 350 mph, an operational range of 4,000 miles and a bomb load of 1000 kg. Initially it looked like the strategic bomber Germany needed, but with Goering's approval, the He 177 had incredibly been assigned to the dive-bomber role. Faced with such difficulties which could not be overcome, Goering had declared to Professor Heinkel:

> "It's absolute rubbish to demand that a four-engined plane should be a dive-bomber. If anyone had told me that that was to be the idea I would have said at once that it wouldn't work."
> "Very well. In that case," Heinkel had replied, "it can very quickly be ready for front-line service."[5]

3 Sqn Ldr S M Green OBE, 262232. 2 AD, No 68 Sqn RAF, 456 Sqn, 3064 SE, 6456 SE. Engineer of Sydney; b 26 Mar 1917.

4 This guided bomb, weighing 500 pounds, was capable of being controlled up to a distance of 2 miles.

5 Bartz, op cit, p.79.

However, when it did enter service, one of the He 177's chief weaknesses was the tendency for its engines to catch fire, and before long, *Luftwaffe* bomber crews were referring to it as the "flying lighter", or the "flying firework". It became much feared, but not by its enemies. In spite of this, production continued, and by 1944 1,100 of them had been built.[6] Of the 26 of this type that were shot down by RAF Mosquitos, 456 Squadron was to be successful in destroying half of them.[7]

June brought with it the expectation of the great day of invasion. The ground crew had been working around the clock on their 24-hour shifts, grabbing sleep whenever they could. Not only was Ford a forward airfield for the invasion, with hundreds of aircraft operating from there, but vibrations of the pattern bombing along the French coast could be felt. LAC Len Stamp recalled:

> As the day came nearer, motor vehicles, trucks, tanks, staff cars, jeeps, personnel carriers, armoured cars were parked on both sides of every town and little village on the entire south coast of England.[8]

On 5 June, the 456 Squadron crews were briefed that the invasion armada would sail across the Channel that night, and they were not to make any mention of the huge number of vessels below them, as they flew their nightly patrols. Fortunately, that night there were no *Luftwaffe* raids, and so the secret of the intended landing location in Normandy was not compromised. On D-Day, 6 June 1944, the Allies had complete control of the air, and flew 6700 sorties. The Germans could only mount 300. The *Luftwaffe's* Air Fleet 3 had only 300 aircraft available, of which 100 were fighters. The Allies had nearly 13,000 aircraft supporting the invasion, of which 5400 were fighters.[9]

That evening was to prove very successful for the Australian night fighters as they patrolled over the invasion beaches. Soon after midnight, Flying Officer Ron Pratt and Flight Lieutenant Stew Smith (HK303) obtained a contact on a He 177 off Normandy. Closing in, Pratt pressed the gun button, only to find his guns failed to fire. After testing them again, they obtained another contact and this time they closed onto a He 177, which they burst into flames. The stricken bomber exploded and crashed into the sea. An hour later, it was the turn of Flying Officer Fred Stevens and Pilot Officer Andy Kellett (HK290). Stevens identified a He 177 carrying two

6 Bekker, op cit, p.230; Bartz, op cit, p.153.

7 See Appendix 8. Sharp & Bowyer, op cit, pp.457-461, provides the total breakdown of Mosquito victories over He 177s as:

25 Sqn RAF	2	29 Sqn RAF	1	85 Sqn RAF	1
151 Sqn RAF	1	264 Sqn RAF	2	307 Sqn RAF	1
409 Sqn RCAF	1	410 Sqn RCAF	2	456 Sqn RAAF	13
488 Sqn RNZAF	1	FIU	1		

8 Len stamp, letter to author, 11 Dec 1994.

9 Bartz, op cit, p.187.

The Mosquito production line at the de Havilland plant at Leavesden in June 1943. These aircraft are NF.II models, but were soon to be fitted with the Mark X AI to become the NF.XVII. Centre left is HK290, behind it is HK286, both aircraft destined for 456 Squadron.

glider bombs, one under each wing outboard of the engines, and with a large aerial array on the nose. After three short bursts, the aircraft caught fire and spun into the sea off Cherbourg. Stevens later described the engagement:

> We were coming in much too fast so I closed the throttles completely and set the range bars to 100 feet. Now when his wings filled the gap we'd be into 250 yards, the best range to open fire. *OK, I've got him* I muttered into the intercom. Andy took his head out of his set to look. I turned the gun button to 'Fire' as we hurtled in the last half mile. He opened fire first, well over 600 yards off. The coloured balls curled lazily away from the rear of the cabin, then accelerated rapidly past our right wing. The dot of the sight settled between the engine and wing root and I pressed the gun button. The armourers had loaded the cannons with incendiary and high-explosive. One of the first incendiaries must have ruptured a fuel cell, for it seemed the moment I pressed the gun button a huge ball of flame engulfed the Heinkel. The crew would have died instantly. The left wing broke off, the rest of the wreckage disintegrating as it fell 10,000 feet into the Channel.[10]

10 Stevens, op cit, p.71.

Flt Lt Stew Smith and Flg Off Ron Pratt

A short time later, Stevens and Kellett obtained a visual on another He 177, and set fire to its port engine. Another burst caused a white flash in the fuselage, and the bomber exploded on impact with the water, off Le Havre. In both these engagements Stevens reported the enemy's use of "Window".[11]

As usual, not to be outdone by the success of 'A' Flight, the CO (HK286) also scored. His navigator, Flight Lieutenant Tom Condon, saw a trail of "window" on his AI scope, and on closing Wing Commander Keith Hampshire gained a long visual at 3,500 feet on a He 177, in magnificent full moon conditions against a white background of clouds. The two glider bombs under the wings were clearly visible. He fired from 150 yards, sending burning debris from the bomber, which then burst into flames, and dropped into the sea east of Barfleur, on the tip of the Cherbourg peninsula.

The following evening, 'B' Flight took on the task. Squadron Leader Bas Howard and Flying Officer Jack Ross (HK323) sent two of the He 177 missile launchers down flaming into the sea off Normandy. They saw a third He 177, but had no ammunition left to throw at it. After a long chase, Pilot Officer "Butch" Hodgen[12] and his RAF navigator, Flight Sergeant A McCormick[13] (HK302), sent another He 177 down in flames, off the invasion beachhead.

---

11 AEAF Oprep A No 68, 456S/3/S308/AIR, 7 Jun 1944, RAAF Historical Section. "Window" was one of the main electronic innovations of the war. Consisting of tin foil cut to the wavelength of a particular radar, "window" ("chaff" to the Americans, "Düppell" to the Germans) was designed to black out an enemy radar screen to hide the presence of aircraft. Window had first been used by Bomber Command on a raid to Hamburg in July 1943, and with its introduction there was a sharp fall in bomber losses. Saunders, op cit, p.29.

12 Flt Lt R D Hodgen, 413594. No 456 Sqn. Clerk of Sydney; b Sydney 28 Apr 1923.

13 WOff A McCormick RAF, 1564851. No 456 Sqn, FIDS.

Two nights later, on 9 June, Flight Lieutenant Bob Cowper and Flying Officer Bill Watson (HK353) were also to score a "double". Watson obtained an AI contact, which enabled Cowper to visually identify a He 177, which opened fire on the Mosquito and took evasive action. Cowper fired three bursts, setting the right wing alight, and he last saw the bomber on fire crossing the French coast at Cap Levy, at about 100 feet. One of its two glider bombs had become detached and caught fire, gliding parallel to the parent aircraft before crashing. Cowper did not see the bomber crash, so the crew were initially only credited with a probable. This claim was upgraded to "destroyed" when the wreckage was later found by the invading armies near Cherbourg.[14] Immediately afterwards, they intercepted a Do 217, flying south around Cap de la Hague. After a 1-second burst from 150 yards, the Dornier, which was carrying a glide bomb under its port wing, burst into flames and crashed on the coast.

Pilot Officer Ivor Sanderson and Flight Sergeant Charles Nicholas (HK353) scored the next night, 10 June, but not without some difficulty. Sanderson became visual with his target over the sea at a range of 2,000 feet, and as he closed the He 177's rear gunner opened fire with his cannon at 800 feet, hitting the Mosquito's starboard wing. This caused major damage to the wing leading edge, and holed the main spar and fuel tank. Sanderson gave two long bursts, and the bomber's starboard engine caught fire, and the aircraft spun into the Channel. The Mosquito was now extremely difficult to control, had no radio, and stalled at 180mph. However, Sanderson was able to return safely to base, and this engagement was cited in his award of the DFC. He later recalled the damage:

> The damaged section of the wing measured between 6 feet and 7 feet and most of the way there was nothing left of the leading edge until the main spar. It was a cannon shell which hit us, we could see the tracer close above the wing. I always approached 10 degrees below and 10 degrees to port. This gave the gunner two deflections to allow for and also allowed for debris coming back to miss us. Some quite big pieces of aircraft seemed to come off when the four cannon struck the target.
>
> When I first fired, I missed completely. I was not a bad shot and didn't realise that the whole plane was flying slightly off true because of the drag to starboard. I fired a circle around the He 177 and a cross through his centre - every possible deflection - this of course got him. It wasn't easy to fly with the stick away over to the port side.[15]

Meanwhile, the 'A' Flight Commander, Squadron Leader Geoff Howitt, and Flight Lieutenant "Red" Irving (HK249) stalked a target over the Channel, about 40 miles south of Brighton, for five minutes, which they identified as another He 177. Firing two bursts into the enemy bomber's port wing and engine, Howitt saw the aircraft dive steeply to the sea, emitting a stream

14 No 456 Sqn Form 540, 7 Jul 1944.

15 Ivor Sanderson, letter to David Vincent, 30 Oct 1984.

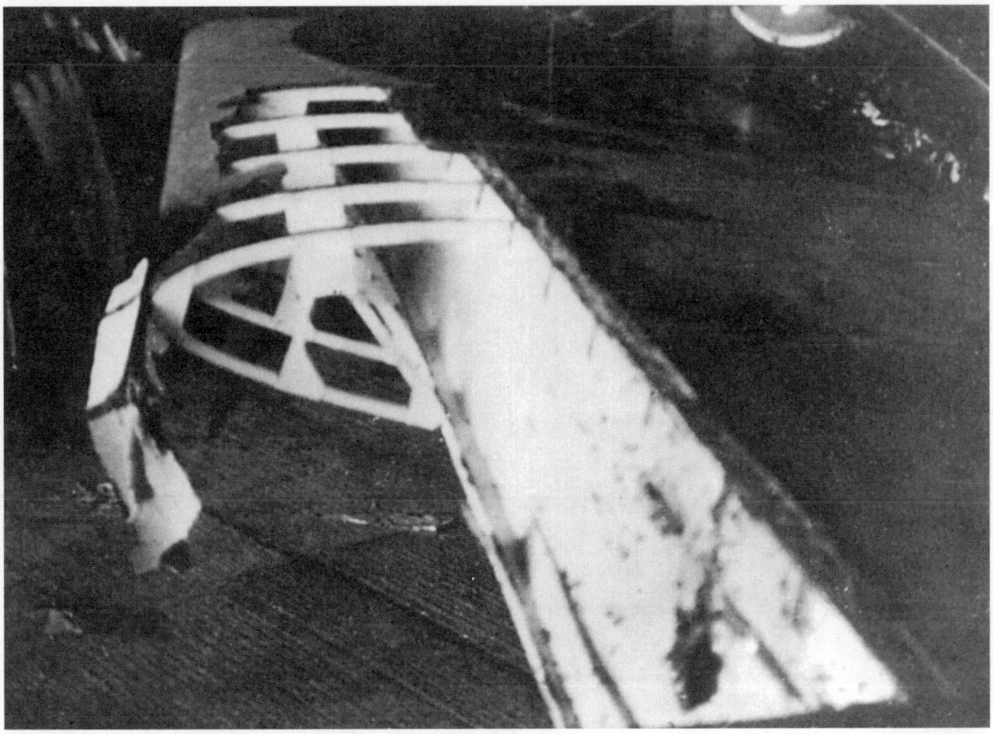

The damaged starboard wing of Ivor Sanderson's HK353.

of white smoke from its burning engine. He lost visual contact with the target, and Irving, who had also had a strong blip on his AI, also saw the target vanish from his scope. No crash, no kill. They had to be satisfied with a probable claim.

The night of 12 June over the English Channel was the CO's turn again. Airborne with Tom Condon (HK286), they obtained a contact at 2 miles, at which time the target took violent evasive action, obviously aware someone was stalking him. Hampshire identified the target as a Ju 88 night fighter, and opened fire at 120 yards. Both engines caught fire, and the enemy aircraft went down out of control, oscillating violently from side to side, finally rolling over vertically, and diving steeply into the sea. German night fighter activity over the Channel had already been experienced that night by Butch Hodgen (HK302), as he had been stalked by two or more Ju 88s. He spotted one just above him, and as he tried to bring his nose onto it, another flashed past, overshooting both him and his prey.

Pilot Officer Stan Williams[16] and Flying Officer Ken Havord[17] (HK282) scored the following night. Finding a He 177 mine-laying and dropping "window" off Fecamp, they quickly despatched the bomber burning into the sea.

16 Flt Lt S J Williams, 409635. No 456 Sqn. Farmer of Swan Hill, Vic; b Swan Hill 25 Dec 1920.

17 Flt Lt K W Havord, 413193. No 151 Sqn RAF, 456 Sqn, 2 OTU. Fitter's labourer of Port Kembla, NSW; b Sydney 23 Dec 1916.

Squadron Leader Geoff Howitt, with 456 fitters Sgt Reg Jeans (left) and FSgt 'Ali' Barber (right). The aircraft is Mosquito NF.XVII, HK249, RX-B.

The next night, 14 June, one of the Royal Navy crews, Lieutenant Dennis Thornley and Sub Lieutenant Derrick Phillips (HK359), had a more unusual engagement. They fired at a target which they had first taken to be a He 177. Their AI ranges and altimeter were unreliable, and they found themselves very close to the sea. On return they found the Mosquito's nose, belly, windscreen and spinners covered with oil, and lodged a claim of "damaged" on their target. From the shape of the wings and nose of the enemy aircraft, it was decided that what the pilot had taken to be a He 177 was presumably a Henschel Hs 129 "tank destroyer"! Flight Lieutenant Bob Cowper and Flying Officer Bill Watson (HK356) had succeeded in a more decisive engagement. They downed a Ju 88 over the Channel near Cherbourg, which caught fire and dived vertically into the sea, with three of the crew baling out on its way down.

Another night fighter was encountered over the English Channel the night of 22 June, when Flying Officer Fred Stevens and Pilot Officer Andy Kellett (HK290) closed on a target, which they identified as a Ju 88G. Firing a burst into the port wing root from 100 yards, the enemy responded with accurate fire from the dorsal turret:

> I fired hurriedly, seeing strikes. A large piece of the wing appeared to blow away after a brilliant white flash. The target, now down to 3,000 feet, rolled onto its back and disappeared.[18]

18 Stevens, op cit, p.73.

GROUNDCREW AT FORD, 1944

Jock Angus and Tony Crooks

Tony Crooks, Ron Smith and Len Stamp
The Oxford aircraft in the background was the station taxi.

As the German dived steeply towards the sea, Stevens was forced to break off his pursuit, as they had been distracted by another AI contact off to the left, which they subsequently lost. Having not being able to keep sight of their original victim, they could only claim a probable.

The hapless He 177s made their final appearance over the Channel on the night of 4 July, and 456 Squadron claimed a further three. Sanderson/Nicholas (HK249), Cowper/Watson (HK356), and Radford/Atkinson (HK312) all scored. In addition, Williams/Havord (HK282) shot down a Do 217. This had been an incredibly successful four weeks for the Australian night fighter squadron. The crews had destroyed 17 enemy aircraft that had been trying to interfere with the invading forces, and had claimed another 2 probables. But now as the Allied armies fought to break out of the beachhead, a new threat had emerged. This was the V1 flying bomb, the "doodle-bug", code-named the "diver".

From left: Sgt Reg Jeans, FSgt Lance Sutherland, FSgt 'Ali' Barber and Sgt Ted Platt.

While 456 had been notching up its aerial victories over the D-Day beaches, other Mosquitos had been engaged in night fighter Rangers. As related by the RAF Official History:

> The Mosquitos were soon not only ranging Normandy, Belgium and France, but also Germany, causing, almost as much by their presence as by their bombs and bullets, a grave check to the flow of reinforcements, without which Rommel's men in the beachhead must perforce perish. Perhaps the most outstanding series of attacks in the early period was that delivered on the night of 7th/8th July when 109 Mosquitos assaulted twenty-six trains and three road convoys.[19]

19 Saunders, op cit, p.119.

HONOURS AND AWARDS [20]

DISTINGUISHED FLYING CROSS

FLYING OFFICER IVOR SANDERSON

CITATION

Flying Officer Sanderson is a keen, skilful and resolute fighter and has destroyed 3 enemy aircraft at night.

In one of these fights his aircraft was badly hit. Nevertheless, he pressed home his attack from close range and shot down the enemy aircraft. His own became difficult to control but he flew it to base and effected a safe landing. He displayed great skill and tenacity throughout.

20 London Gazette No 36931, 9 Feb 1945.

CHAPTER FOURTEEN

ANTI-DIVERS

As a substitute for their lack of a strategic bomber, the Germans had begun development of a range of revenge weapons. The V1 was a pilotless flying bomb powered by a pulse jet and launched from a ramp, which carried a 2,000-pound warhead an average range of 250 miles. The V1 program had been hastily conceived, and flight testing at the Peenemünde research facility in northern Germany over 1943 had been largely unsuccessful. By December, the accuracy and reliability of the missiles were still so poor that only one out of every six launched from their ramps would have hit London.[1] In addition, production in Germany was far from proceeding satisfactorily, as Allied bombing attacks on the V1 factory had caused quite a delay to the program. The bombing of the Fieseler Works at Kassel, in October 1943, halted supplies of flying bombs to Peenemünde until February 1944.

"Ski" sites for the V1 onslaught, so-called after their distinctive ski-shaped launch ramps, had been constructed along the French coast. Forming an arc between Dunkirk and Abbeville, the ramps were all aimed at London. As an active countermeasure, Allied bombing of the ski sites, which were code-named "No-ball" sites, had begun on 21 December 1943, with No 464 Squadron RAAF, flying Mosquitos, joining operations the following month. By April 1944, the Allies felt that all 96 launch sites were practically neutralised, and by the end of May, 103 of the 140 sites had been destroyed.[2]

The Allied No-ball offensive forced the Germans to modify their

1 Jones, op cit, p.467.

2 Jones, op cit, p.525; Saunders, op cit, p.152.

deployment program. Not only had the bombing of the sites been so successful in delaying the start of the V1 attacks, but it also forced the Germans to use new degraded sites. These consisted of a simpler, prefabricated ramp, which could be more rapidly erected, and the modified sites were camouflaged to resemble farm buildings. First observed in April 1944, the new sites were being constructed faster than the Allies could destroy them. Even though these sites were not being destroyed, by the Germans adopting the modified sites undoubtedly contributed to a degradation in accuracy for when the onslaught began.

British Intelligence had previously assessed that commencement of the V1 campaign would be provoked by the D-Day landings, and had estimated that it would take a week for the Germans to be able to react. Consequently, D+7 had been the best guess for the opening of the bombardment.[3] True to this accurate assessment, on 13 June 1944, the first V1s, or "divers", were launched against England. On this first day of the V1 campaign, of the 55 sites technically ready, seven alone fired, launching ten bombs. Four bombs crashed not far from the site, and three detonated in the air. Only four of the ten actually reached England – one landed at Sevenoaks in Kent, another near Gravesend, a third at Cuckfield, Sussex, and the fourth at Bethnal Green, London.

The following night an RAF Mosquito crew were over the Channel when they spotted a "ball of fire" approaching from the Continent. The pilot described their encounter:

> It flashed past on our starboard side a few thousand feet away and at the same height as we were flying. I quickly turned to port and chased it. It was going pretty fast, but I caught up with it and opened fire from astern.[4]

After three bursts from the Mosquito's cannon, the diver exploded, crashing vertically into the sea. This was, in all probability, the first RAF victory over a flying bomb, and it was a new role for the night fighter squadrons. It was the fighters that played the leading part in the successful combating of the V1 threat, by maintaining the anti-diver patrol line out in the Channel, beyond the anti-aircraft gun belt. By day, it was the job of the Tempests, Typhoons and Spitfires. By night, it was the job of the Mosquitos.

The Germans had soon overcome their teething problems with launching and guiding the V1s. On 16 June, 244 bombs were launched from the 55 sites, and by 21 June, the 1,000th bomb had been despatched against London. The 2,000th was launched the following week. The threat of the flying bomb forced Bomber Command to deploy 30% of its force against the V1 launch sites. V1 attacks on England were now a daily occurrence, coming at a daily average of about 190.

---

3 Jones, op cit, p.482.

4 Saunders, op cit, p.142.

The V1, a small fast target

By the end of June, two methods of controlling the fighters had been devised to counter the V1. The first was under the close control of the GCI radar stations along the coast. As soon as a V1 appeared on the GCI radar scope, the patrolling fighters were directed to intercept. The technique was to fly some 5,000 feet above and across their path until their flames were visible, when a peel-off and full-throttle dive might catch them.[5] The problem was there were only four GCI sites on the south-eastern coast, and they could not detect the bomb beyond fifty miles.[6] This gave the fighter six minutes to intercept and shoot it down. If this could be accomplished, the bomb fell harmlessly into the sea. The other system of control was the "running commentary" over the land. The position and course of the bomb was passed by ground control to all fighters working on the same radio frequency. The fighter then calculated its own course to achieve an intercept.

To acquire the bomb involved looking down into the blackness, searching for the moving yellow light of the long exhaust flame, followed by a dive in from a rear quarter at 320 mph, switching on the aircraft's navigation lights

5 Meadows, Part 4, p.45.

6 Saunders, op cit, p.161.

and keeping an eye out for any other aircraft. Quite often, more than one fighter went for the one bomb. The small size of the flying bomb made interception difficult, and the margin of superiority in speed was small. The pulse jet gave the V1 a speed of between 250 and 400 mph, flying between 1,000 and 5,000 feet. Therefore, a stern chase was usually futile, so the fighter would try to start above the bomb to obtain a speed advantage in the dive. The interception had to be made very quickly, before the bomb reached the gun and balloon belts in front of London.

Having intercepted the diver, the next problem was to shoot it down. The ammunition loaded by the armourers for these missions now included tracer rounds. Tracer had not normally been fired by night fighters - it gave up that element of surprise - but against the V1, a mix of one tracer in every four cannon shells proved useful in gauging the range against these tricky targets. It was necessary to get within 200 or 300 yards of the flying bomb to be assured of a good shot.

One successful tactic was to fly a parallel course ahead of the bomb, firing deflection bursts as it passed. It was necessary to remain outside 100 yards when firing - the explosion of the bomb would destroy the fighter. Several unorthodox tactics were also tried by pilots. In daylight, the stubby wings of the V1s were tipped by fighters' wingtips, upsetting the bombs' guidance, or the V1 could be thrown into a spin by an aircraft's slipstream. However, such manoeuvres were hazardous and not encouraged, especially at night.

To improve the performance of the Mosquitos for the anti-diver operations, several modifications were incorporated. The Merlins' exhaust flame shrouds were removed, and engines were boosted to provide extra power, and were now run on 150 octane fuel. In addition, the nose was strengthened to withstand the effects of cannon fire at these high speeds. The patrols were flown at 8,000', above the divers flying typically at 2,000'.

By July 1944, the V1 flying bomb assault was getting into full swing by day and night. ADGB made every effort to thwart the V1s from reaching civilian targets, and night fighters were despatched on watching patrols along the French coast in the hope of catching the bombs before they reached England. Three main forms of defence were formed - AA guns, balloon cables, and RAF fighters. The AA defensive belt built up to 620 heavy guns and 1,762 light guns.[7] The guns were moved during mid-July forward to the coast, a massive deployment of 800 guns, 60,000 tons of stores and ammunition, and 23,000 personnel. The anti-diver forces were deployed by 17 July in four distinct belts:

Fighter Belt at Sea. Fighter aircraft operating under close radio control, at a distance not less than 10,000 yards from the shore.

Coastal Gun Belt. Anti-aircraft guns allotted for defence against the V1 deployed in a narrow strip 5,000 yards wide, extending along the coast

7 Franks, op cit, p.194.

from Beachy Head to St Margaret's Bay, firing out to sea up to the 10,000 yard limit.

Inland Fighter Belt. A second fighter belt operating under running-commentary control. The bursts of anti-aircraft fire from the coastal gun belt indicated the flight path of any approaching bombs. There was additional assistance at night from searchlights over the whole of the inland fighter zone.

Balloon Belt. There was then the existing balloon barrage. [8]

Within two days of the move to the coast, the guns were being more successful. The gun scores now rose from 50 bombs a week to 170 a week. The fighters were now more restricted, and their score dropped from 180 to 120 a week, but the overall gain was substantial.[9]

No 456 Squadron had first engaged divers during its nightly Channel patrols during the latter part of June. The Naval crew, Thornley and Phillips (HK359), had unsuccessfully engaged a diver the night of 17 June, and five nights later Fred Stevens and Andy Kellett (HK290) had fired at another, only to find the V1 draw rapidly away. As Stevens later stated: "These targets were so tiny!"[10] On the 24th, George Houston and Gus Engberg (HK264) made several attempts at intercepting divers from their patrol line east of Beachy Head. On one, the diver was fired at by another fighter which swooped upon it. Their next attempt was more productive, scoring strikes on the bomb, which appeared unaffected, continuing on straight and level. Three minutes later an explosion was seen on the ground, about 18 miles inland. The first diver to 456 - a probable.

The Australians were now tasked almost exclusively on anti-diver operations during July, by which stage practically all the Mosquito night fighter squadrons were involved on these patrols.[11] The mission was clear - destroy the V1 before it reached the AA belt when patrolling over the sea, or before it reached the balloon belt when patrolling inland. The first success came on the night of 9 July, when Flight Lieutenants Keith Roediger and Bob Dobson (HK297) engaged a V1 flying at 2,500 feet on a north-westerly heading. Roediger closed to 500 yards, offset himself to the right to avoid the debris, and shot it down into the sea The next night, Houston/Engberg (HK264) shot down another. Killed during these operations in the early hours of 12 July, were Flying Officer Ted Radford and Flight Sergeant Wally Atkinson (HK312), who crashed in the sea 5 miles south-east of Littlehampton. By the end of July, 456 had claimed ten divers destroyed, three out to sea and seven inland (see Appendix 9).

8 W S Churchill, *The Second World War, Vol VI*, Cassell, London, 1954, p.42.

9 Jones, op cit, pp.540-1.

10 Stevens, op cit, p.75.

11 These were Nos 25, 68, 85, 96, 125, 219, 264, 409 and 456 Sqns, joined by 157 Sqn in August, and 29 Sqn in October.

Mosquito NF.XVII HK312 RX-G, lost off Ford with Ted Radford and Wally Atkinson.

456 Squadron's normal nightly "beat" on the inland fighter belt was the 50 mile patrol line from Beachy Head to Bognor Regis. Every ten miles was a master searchlight, which shone vertically and flashed a morse-coded identification: R -Robert, S - Sugar, T - Tare, U - Uncle, and V - Victor. These enabled the anti-diver patrols to keep to their allocated positions. The five mile strip either side of the searchlight was designated Robert West and Robert East, then Sugar West, Sugar East and so on. On one occasion the CO had just returned from leave and, during his absence, abbreviated instructions had been introduced by the GCI controllers. When getting airborne on his first night back, he was directed "Tare West". After some fifteen minutes, the GCI controller heard a faint transmission: "Approaching Land's End, and still tearing west!"

The anti-diver defences were at their greatest strength by mid-August, with 21 day and night fighter squadrons of Air Defence of Great Britain engaged entirely on diver patrols.[12] The scoring for 456 for August was opened by Keith Roediger and Bob Dobson (HK297) on the night of the 4th, followed the next night by Flying Officers John Semmens[13] and Hedley Nitschke[14] (HK253). By the end of the month the score stood at 29, with

12 In addition, there were over 2000 balloons, 1500 guns and 600 rocket firing barrels. Saunders, op cit, p.167.

13 Flg Off J E Semmens, 412849. No 456 Sqn, 667 & 285 Sqns RAF. Farmer and grazier of Leeton, NSW; b Lockhart, NSW, 9 May 1918.

14 Flg Off H A Nitschke, 417222. No 456 Sqn, 51 OTU. Shop assistant of Loxton, SA; b Saddleworth, SA, 27 Apr 1912.

The top scorers on 456 against the V1 flying bombs, Keith Roediger and Bob Dobson.

Roediger/Dobson as the most successful crew, destroying nine of the flying bombs.[15]

The record "bag" of divers for ADGB was on August 28, when 94 V1s approached the coast and all bar four were destroyed. The balloons scored two, the fighters 23, and the guns 65.[16] Of the 1,124 bombs fired over the last three weeks of the campaign up until 5 September, only 17% fell in the target area. By this stage, as the Allied armies advanced in France and overran the sites, it appeared that the threat from the V1 was concluded.

Now the Mark X AI was cleared to overfly occupied territory, and the Australians spent the latter half of September, all of October, and into November flying from the forward base at Manston, Kent, either providing night cover for the Arnhem airborne operation or flying uneventful patrols over the stabilised fronts in Belgium and Holland. A GCI station, known as "Rejoice", had been set up near Brussels to control night fighter activity over the Continent, and this site controlled 456's solitary success on these operations during the night of 6 October. Warrant Officer John Mulhall[17] and

15 Some sources credit 456 Sqn with a total of 24 V1s destroyed. The 456 Squadron Form 540 (Operational Record Book), for July and August 1944, is quite clear that 29 were destroyed.

16 Churchill, Vol VI, op cit, p.43.

17 WOff J L Mulhall, 410361. No 456 Sqn. Clerk of Melbourne; b Melbourne 19 Jul 1923. Killed in action 24 Nov 1944.

his RAF observer, Flying Officer "Taffy" Jones[18] (HK317), were vectored onto a target, which they identified by its sharply tapering wings and squarish fuselage as a Ju 188. Firing into its starboard engine, the bomber was sent down in flames, 20 miles north-west of Nijmegen. The crew, and that of another Mosquito, Flying Officers Dick Wilkie[19] and Scotty Fraser (HK313), saw the aircraft explode on impact with the ground.

On the night of 6/7 November, one of the Squadron's crews failed to return from a Continental patrol. Lieutenant "Woodie" Woodward and Ensign "Jerry" Madden were one of two US Marine Corps aircrew who had been attached to 456 Squadron for operational night flying experience before returning to the US to instruct on night fighting techniques.[20] Flying aircraft HK357, they departed for a defensive patrol over Holland and Belgium, and after their departure call to Durrington GCI, their fate was unknown. That night, four night fighters from the Squadron were tasked on defensive patrols to deter any enemy air activity that might interfere with the Allied armies, that were experiencing reversals around the Dutch-German border:

> We flew [HK356] in the hideous storm to Arnhem to help cover the retreat...the wind was gusting at 100mph in the Channel, and indeed Brighton pier was blown down. It was the worst night's flying we had ever experienced...[21]

This view was also shared by Watson's pilot, Bob Cowper.[22] It had been a sorry experience for the US Marines, as a total of five crews were attached to the RAF to fly Mosquitos. The following week, on 14 November, an American crew flying with 68 Squadron was shot down by British guns as it pursued an air-launched V1 into the defensive gun belt protecting the eastern coast. The crew's enthusiasm had evidently got the better of them as they chased the V1 into the guns, instead of pursuing the mother plane. A memorial to the dead crew, Lieutenants J F Black and T N Aiken, was erected close to the crash site at Decoy Farm, Blundeston, Suffolk.[23] In all, only two out of the five USMC Mosquito crews survived. The remaining Americans from 456, Archie Sinclair and "J" Gould returned to the US, and in May 1945 organised for the presentation of a silver bowl to the Squadron to commemorate the American association with the RAAF. The bowl was inscribed:

---

18 Flg Off J D Jones RAF, 53112. No 456 Sqn. Killed in action 24 Nov 1944.

19 Flt Lt R J Wilkie, 417140. No 456 Sqn, 51 OTU, 25 Sqn RAF. Bank clerk of Terowie, SA; b Adelaide 25 Oct 1921. Killed in flying accident 7 Jun 1949.

20 The two USMC crews were Lt E M Woodward/Ensign W G Madden and LCdr A R Sinclair/ LCdr J H P Gould.

21 Dr W Watson, letter to David Vincent, 17 Sep 1985.

22 Bob Cowper, letter to author, 29 Nov 1994.

23 Mosquito HK289 (WM-K of 68 Sqn). Flypast magazine, Key Publishing, Stamford, Lincs, December 1982; and D J Smith, De Havilland Mosquito Crash Log, Midland Counties, Earl Shilton, Leics, 1980, p.25.

With Appreciation
From the four US Navy types
Who are proud to have served with you.

"J" Gould	"Woodie" Woodward
Archie Sinclair	Jerry Madden[24]

A novel was subsequently written about five US Navy and Marine crews who were attached to fly with RAF Mosquito night fighter squadrons.[25] It describes two crews flying with 456 Squadron, and three with 68 Squadron, and how three of these crews failed to return in November 1944. It then relates the execution of a Ranger mission against Goering's personal train. This fiction is clearly based upon the attachment of the Americans crews, and it is acknowledged that the author, "John Kelly", himself had been a night fighter pilot with the US Navy and RAF during the war. Could he be one of the two surviving Mosquito pilots, perhaps even 456's Lieutenant Commander Archie Sinclair?

Sinclair and Gould left 456 Squadron on 1 December 1944, returning to US Marine flying duties in the States. The inscribed silver bowl has resided since the war with 456's last CO, Squadron Leader Bob Cowper, and it serves as the Squadron Silver for 456's annual reunions.

The Germans had by now further developed the V1 to enable it to be launched by an aircraft. From July 1944, V1s were air-launched from Heinkel He 111H twin-engined bombers, flying with III/KG3 from bases at Venlo and Gilze-Rijen, in Holland. These attacks were launched from the North Sea, with the intended targets being London, Portsmouth and Southampton, outflanking the anti-diver defensive belt in Kent, which had been designed to counter attacks from France. During September, by which time some 400 V1s had been air launched from the Heinkel 111s, III/KG3 moved its bases from Holland to north-west Germany, continuing V1 operations as a new *Gruppe*, I/KG53.[26]

These aircraft, each capable of carrying two V1s under the wings, would fly at 300 feet over the North Sea to avoid the night fighters, popping up to about 1,500 feet when about 60 miles from the English coast. The Heinkels would increase speed to the bomb's launch speed of around 200mph, while the flight engineer started the V1's motor.[27] The missile would be released at a range of between 30 and 40 miles, with the launch aircraft on a direct bearing to the intended target. The Heinkel would then dive back towards the sea, for a hurried return to home base. In practice, accuracy suffered from these air launches, as both the course and distance had to be pre-programmed into the V1 before the launch aircraft took off, allowing no error in navigation to the release point.

24 No 456 Sqn 540, 7 May 1945.

25 J W Kelly, *The Wooden Wolf*, Futura, London, 1978.

26 F H Hinsley, et al, *British Intelligence in the Second World War, Vol 3, Part II*, HMSO, London, 1988, pp.545-8

27 Ramsey, op cit, p.412.

The air-launched flying bomb - a Heinkel He 111 releases its V1.

456 Squadron was back at full strength at Ford when anti-diver patrols started again on 5 November against the He 111 aerial launchers. The first success came a fortnight later, on the night on 19 November, when Flying Officers Doug Arnold and John Stickley (HK246) destroyed a He 111H V1 launcher. Under the control of Trimley GCI, on the East Anglian coast near Felixstowe, the Mosquito crew were at 300 feet over the water when they gained a visual contact on the He 111, just after it had fired a V1. Arnold chased the target and closed to almost 100 yards, firing three bursts, and saw the bomber return fire, which damaged the Mosquito's starboard propeller blade. However, his strikes had been accurate. The Heinkel's starboard engine began burning, and then the bomber crashed into the water, 75 miles east of Lowestoft, in the middle of the North Sea. Arnold and Stickley brought their damaged fighter back safely to base.

To enable flying at such low altitudes at night, the Mosquito pilots were learning to rely on a new instrument, the radio altimeter, which provided an accurate height above the ground, or in these operations, above the water. The height indicator lights were on the right side of the instrument panel. When set to the desired level of 250 feet above the water, coloured lights provided height information. A white light indicated when above 300 feet, a green showed the desired height band between 200 feet to 300 feet, and a red light warned when the aircraft descended below 200 feet.[28]

It was in this system that Flying Officers Fred Stevens and Andy Kellett (HK290) were placing their trust, when they flew their patrol off Norfolk

28 Fred Stevens, interview with author, 11 Dec 1994.

during the early hours of 25 November. It was pitch dark as they set up their patrol line at 1,000 feet, flying a race-track pattern to the south-east under the control of Bawdsey GCI, on the Suffolk coast. As Stevens recollected, soon Bawdsey called:

> "Larkin 26[29], this is Frogspawn, you have a contact very close to you!" Andy's head was in the AI scope: "Contact right below us."
>
> So began a 25 minute chase, the target making continuous weaves, altering heading and height, forcing us to get our speed down to 140mph, with flaps and wheels down. We did an orbit to let the enemy aircraft get ahead. As the target climbed to 600 feet and straightened on a north-easterly heading, into a lighter part of the sky, we were able to identify it as a Heinkel. As we closed, the rear gunner opened fire and I pressed the button, firing into the cabin area. Our high closure rate meant I had to yank the wing over his fin, and continued into a steep left turn to position for a second pass. But the damage had been done on the first attack, and the flaming He 111 dived into water. A flash lit up the sky as the burning wreckage bounced over the surface leaving a sea of flames.[30]

It had been a long chase, the German almost making it to the Dutch coast, before going down, 10 miles east of the island of Texel. Fred Stevens described it as: "the most exhausting 25 minutes of flying I've ever had".[31]

The two victories against the He 111 V1 launchers had not been without cost. The mission of trying to line up a Heinkel in pitch blackness, at only a few hundred feet above the water, while turning at very slow speed, was a demanding one. In the early hours of 24 November, a Mosquito crewed by Warrant Officer John Mulhall and Flying Officer Taffy Jones (HK317), failed to return from such a sortie:

> Chasing the He 111 with their buzz bombs was the toughest flying test ever asked of us. They flew below our stalling speed, only feet off the water. Even after a lecture from me the night before, when I was Flight Commander of 'B' Flight, Mulhall and his observer killed themselves trying too hard.[32]

In four months, 41 He 111s were lost on operations, 77 if counting those lost on training accidents. Mosquitos claimed 16 Heinkels destroyed on these operations. Over the entire period of the V1 onslaught, it has been estimated 7,074 bombs had approached near enough to Britain to be

29 For operations from Ford, 456 Sqn used the callsign "Larkin": known individual callsigns were Larkin 15 McLennan; 18 - Darling; 19 - Pratt, then Darling; 25 - Walters; 26 - Stevens; 27 -Cowper; 28 - Griffin; 38 - Roediger; 41 - Hodgen.

30 Fred Stevens, interview with author, 11 Dec 1994.

31 Fred Stevens, interview with author, 11 Dec 1994.

32 Bob Cowper, letter to author, 29 Nov 1994.

reported to the defences, of which 5,377 crossed the coast. In all, 8,893 were launched from the ramps, 1,660 from bombers. Of the V1s destroyed, 1,846 were shot down by fighters, 1,867 by AA fire with another 232 brought down by barrage balloons. More than 6,000 people in Britain died as a result of V1 attacks.[33]

With the ending of V1 campaign, Fighter Command, as it was once again known, was able to increase the number of sorties in support of Bomber Command operations. Wing Commander Hampshire had been promoted and posted to Transport Command, and had been replaced by his 'B' Flight Commander, Bas Howard, on 10 November. Keith Hampshire had served as CO for almost a year - the longest serving commander of 456. Geoff Howitt was 'A' Flight Commander and Bob Cowper had taken over 'B' Flight, as bomber support operations became the next task for the members of 456, under Wing Commander Howard.

33 Franks, op cit, pp.197-8; Saunders, op cit, p.169.

HONOURS AND AWARDS[34]

BAR TO THE DISTINGUISHED SERVICE ORDER

GROUP CAPTAIN KEITH HAMPSHIRE

CITATION

Since being awarded the Distinguished Flying Cross this officer has led the Squadron on very many sorties during which 32 enemy aircraft have been destroyed. Much of the success achieved can be attributed to this officer's outstanding leadership and great skill. His iron determination and unfailing devotion to duty have set an example of the highest order. Group Captain Hampshire has inflicted much loss on the enemy including the destruction of seven aircraft.

34 London Gazette No 36931, 9 Feb 1945.

CHAPTER 15

BOMBER SUPPORT

Bomber Command, which was confident it had won control of the air over Europe by November 1943, had again begun suffering heavily by December. These losses continued over winter, reaching an all-time high in March 1944. German night fighters achieved such success largely due to their *schräge Musik* (literally slanting music) twin 20mm obliquely upward-firing machine guns, that Bomber Command was compelled to draw away from its primary target of Berlin. High losses had occurred on the night of 19/20 February 1944 when 78 of 823 bombers were lost, and on 24/25 March 1944, when 72 bombers failed to return. The following week, on 30/31 March, out of 795 bombers attacking Nuremberg, 95 were lost and 71 severely damaged, with 12 being written-off on landing. This was an incredible loss rate of 12%, causing the night air offensive against Germany to be temporarily suspended - the end of the Battle of Berlin.[1]

Over Germany the RAF resumed night bombardment in late April 1944, and by autumn had won full control of the air by the new tactics and jamming devices which finally even put the night fighters' *Lichtenstein* SN2 AI radars out of action. For bomber support, 100 Group aircraft were filled with all types of radio and radar equipment to mislead ("spoof") and jam German radar screens and radio communications. These were Bomber Command's radio countermeasures experts, whose role was to frustrate the German night defences. In addition, there were the bomber support fighters, to destroy enemy fighters before they could infiltrate and hide within the bomber stream.

Bomber support operations had initially involved five Mosquito night

1 Bekker, op cit, pp.339-40.

fighter squadrons (Nos 23, 141, 169, 239 and 515 Squadrons) as part of 100 Group. But the mounting bomber losses had caused the transfer during May 1944 of two more units, Nos 85 and 157 Squadrons. Although being diverted to anti-diver missions over the summer and autumn, AI Mark X squadrons returned to bomber support duties in September. As enemy forces were rolled back on the Continent, so the air activity over Britain decreased and the night fighter force could be depleted. By the end of 1944, this permitted the release of more of Fighter Command's Mosquito XXX units (Nos 151, 307 and 406 Squadrons) from defensive to offensive tasks, mounting standing patrols over enemy airfields while bomber raids were in progress to pin down the night fighters. No 456 Squadron was to follow with this tasking in March 1945.

456 Squadron had already flown Mahmoud sorties on bomber support during 1943, but Mahmoud was, from the outset, only considered a temporary measure. Now better equipment was available. Missions involved maintaining radio silence until crossing the 5 degree east meridian (approximately the western border of Germany), with the operations being more sophisticated and coordinated, and planned in three phases. Firstly, low level intruders, carrying bombs, would patrol the night fighter airfields. Next, AI Mark X equipped Mosquitos flew in low, and then climbed ahead of the bomber stream to pick off the first enemy fighters, then continued patrolling until about 30 minutes after the raid. Finally, low level intruders with bombs patrolled the airfields awaiting landing night fighters.[2]

A concern of the RAF at the end of 1944 was the possibility of German night fighters operating large-scale intrusions over England to attack returning bombers. The breaking of the Enigma codes had suggested that this was in fact being planned, and Fighter and Bomber Commands were prepared.[3] However, the *Luftwaffe* did not have enough strength left to mount such a campaign, and instead tried to break the tactical air forces on the Continent with a final assault on Allied airfields on the morning of 1 January 1945.

During early December, 456 Squadron had been advised it would re-equip with the latest Mosquito, the Mark XXX. The Mosquito XXX, soon designated the NF.30, represented the high point of wartime night fighter development, retaining the AI Mark X radar of the Mosquito XVII. This new model was powered by the two-stage Merlin 70 series engines with paddle-bladed propellers,[4] which gave a top speed of 424mph at 26,500 feet, 338mph at sea level, and an operational ceiling was 35,000 feet. The range at cruising speed was 1,180 miles, when flying at 30,000 feet. It was, however, a heavy aircraft, with an approach speed of 150mph, coming over the end of the runway at 130mph. Having first flown in March 1944, by the end of the year seven squadrons had NF.30s, which was the most numerous of the late war Mosquito night fighters, with 530 being built.

2 Sharp & Bowyer, op cit, p.292.

3 Jones, op cit, p.591.

4 The Mosquito NF.30 was powered by the 1680hp Merlin 72 or 1710hp Merlin 76 engine, with a two-stage supercharger for better high altitude performance.

Mosquito NF.30 NT296 RX-S. From left: Tony Masters, Bill Johnston and Maurie Quinlan.

The first three NF.30s for 456 were delivered from No 96 Squadron at Odiham, Hampshire, on 7 December, as that unit was disbanding. The NF.30 was also equipped with the 'Gee' grid navigation system, the name Gee being derived from the letter G for grid (see Appendix 10). Gee was relatively easy for the navigator to operate and he could obtain a fix of the aircraft's position fairly quickly within the system's coverage over the UK and the Continent. By late 1944, the Germans had begun jamming the system, however, this interference was not to prove very effective and Mosquito crews were still able to obtain Gee fixes.

The first operation with the new NF.30 was flown the night of 12 December, by Wing Commander Bas Howard and Flight Lieutenant Jack Ross (NT264) on an anti-Heinkel patrol over the North Sea. Airborne at the same time on a similar patrol was Flight Lieutenant Tony Samson and Flying Officer Alec Abbey (HK264), who had numerous contacts on targets and finally closed to identify a strange looking twin-engined bomber. Having difficulty identifying the aircraft, Samson flew several orbits to discover that it was a Wellington, which then opened fire on the Mosquito. The night fighter took serious damage to the starboard engine and wing, but was able to limp back to Woodbridge, Essex, on one engine.

The end of the year brought the news of postings and another relocation, and it was news received with deep regret as Ford had been home for the past ten months. Squadron Leader Geoff Howitt was posted to command No 125 Squadron, and he was replaced as 'A' Flight Commander by Bob Cowper who, in turn, handed over 'B' Flight to Squadron Leader Tony Samson. On 30 December the Squadron moved north to the 12 Group station at Church Fenton, in Yorkshire, to replace No 307 Squadron, a Polish night fighter unit. It was a big move, being quite a distance north to Leeds, and owing to the changeover from the older Mosquitos to the NF.30, meant

The white landscape of Church Fenton in winter.

that 456 had 27 aircraft on strength. Church Fenton was a small peacetime station, with suitable buildings and facilities for maintenance, but with basic and crowded domestic accommodation. The Officers' Mess was to become known as "Bleak House"! Here the Australians would endure their coldest winter of the war.

The weather, and the training required with the Gee nav system, limited operations through January. February saw training continuing, together with uneventful patrolling off the east coast, as the new aircraft were now fitted with the Monica tail warning radar in preparation for bomber support patrols over the Continent, as Fighter Command returned to the bomber support role to reinforce 100 Group. On one of these exercises on 14 February, Squadron Leader Bob Cowper and Flight Lieutenant Bill Watson (NT299) were on a cross-country navigation exercise at 15,000 feet, when they experienced an engine problem. They were directed to land at Woodvale, south of Blackpool, but they overran the short runway, wrecking the aircraft, which ended up on an adjoining golf course, the wheels being ripped off as they had gone through the bunkers. Bob Cowper recalled:

> Woodvale control tower then said "OK to taxy back, Larkin", just as an airman jumped the airfield perimeter fence and ran up. He said "You're on the golf course", to which Bill, a keen golfer, replied "And I haven't brought my clubs!".[5]

After these two months of relative quiet at Church Fenton, the Squadron

5 Bob Cowper, interview with author, 21 Dec 1994.

was advised that it would move on 3 March to the Essex coast base of Bradwell Bay, to operate with No 151 Squadron on full-time bomber support duties. However, this move was to proceed very differently from anything that had been experienced before. For the first time since 1941, the *Luftwaffe* night fighter force conducted intruder attacks against RAF airfields. On the night of 3 March, 140 German night fighters took off to attack Bomber Command airfields ranging from Northumberland to Oxfordshire. Although less than 100 succeeded in penetrating overland, 22 RAF bombers were shot down over their bases.[6]

'A' Flight Commander, Bob Cowper

Being on stand-down for the impending move to Bradwell Bay, with no groundcrew to see them off, it said a good deal that 456 Squadron were able to respond to the German raid. Even the popular squadron doctor, Wing Commander Frank Parle[7], was acting in the capacity of a fitter to see the fighters off. One navigator, Bill Watson, recalled: "A Black Comedy if ever there was one."[8] His pilot, the 'A' Flight Commander, Squadron Leader Bob Cowper, recorded in his logbook "ops scrambled from party, no ground crew, no nothing!", and remembered: "...all landed safely. Phew!"[9]

6 Air Ministry Pamphlet No 248, p.383.

7 Wg Cdr F S Parle, 261215. SHQ Bankstown, 7SFTS, No 456 Sqn, 2PD. Doctor of Sydney; b 22 Jun 1915.

8 Dr W Watson, letter to David Vincent, 17 Sep 1985.

9 Bob Cowper, letter to author, 29 Nov 1994.

A follow-up attack on the following evening, 4 March, was not so successful, with 20 German aircraft being shot down. Five 456 aircraft were scrambled, with Flying Officer Butch Hodgen and Warrant Officer McCormick (MM687) being directed by Orby GCI onto a Ju 188, 30 miles east of Spurn Head. Hodgen fired into its port engine and wing, when suddenly his gunsight containing the lamp fell of. The light dazzled him and he lost his visual. Further searching proved unavailing, and the crew could only claim a "damaged". Flight Lieutenants Harry Smith and Doug Wiley[10] (NT300) closed on a target which they identified as a Ju 188, but their overtake meant they had to break away. Closing again, the bomber dived violently down towards the sea. Smith chased the bomber for over half an hour at only 150 feet over the water before having to give up the chase.

After this enemy activity, 12 Group directed the Mosquitos to remain at Church Fenton. Fifty groundcrew were flown back by Dakota and the train with the Squadron's equipment was recalled from Bradwell. A state of ten aircraft was required, however, these forays were never again repeated by the *Luftwaffe*. After some uneventful patrols, the order to move was revived, and 456 commenced operating from Bradwell Bay on 17 March.

From left: Harry Smith, Jock Donald, Bob Lewis and Hoppy Williams.

Bradwell Bay, on the Essex coast near Southend-on-Sea, had become operational in the Hornchurch Sector in April 1942, and the base currently

10 Flt Lt D J Wiley, 403703. No 604 Sqn RAF, 54 OTU, 456 Sqn. Clerk of Sydney; b Brisbane 4 Oct 1918.

NT295 RX-P in servicing, around April 1945 at Bradwell Bay.

was home for the Mosquitos of No 151 Squadron, which had recently switched from V1 interception to bomber support duties. The station was considered an ideal base from the operational point of view, with excellent runways, comfortable dispersals and good general facilities, however, it was rather inaccessible on the mud flats of the estuary of the Blackwater River.

The two Mosquito squadrons now constituted the "Bradwell Wing". The operations from now until the end of the war were part of the general activities of No 100 Group They were either high altitude sorties, normally flown at 18,000 feet, or low level patrol, penetrating deeply into Germany or neighbouring areas, and being at extreme range, required the fitting of long range tanks to the Mosquitos. The low level sorties were virtually Rangers, but the prime targets were airfields, with secondary targets of opportunity, such as road transport and trains.

Bomber support operations were commenced on the night of 27 March, when the new role was inaugurated by the CO and the two Flight Commanders, who flew Ranger sorties over Germany with varying fortunes. Wing Commander Howard and Flight Lieutenant Jack Ross (NT264) set off for Ingolstadt, near Munich, but after almost two hours had to turn back with an engine problem. Squadron Leader Bob Cowper and Flight Lieutenant Bill Watson (NT300) patrolled Celle and Alzwedel, near Hannover, and had one uneventful chase when their target drew away. Squadron Leader Tony Samson and Flying Officer Alec Abbey (NT282) patrolled Halle, near Leipzig, and shot up some motor transport after seeing the headlights on the road.

On the night of the 30th, six patrols were flown over a wide area by three crews from each flight. The CO and Jack Ross (NT264) patrolled the

Ingolstadt area, and destroyed four MT vehicles. Flight Lieutenants George Houston and Tom Condon (NT241) patrolled west of Berlin and destroyed a train, while Flying Officer Geoff Rees[11] and Pilot Officer Ron Letten[12] (NT296), on their first offensive sortie, destroyed a further two trains in the same area.

Train-busting Ranger sorties started off April, as patrols in support of the bombers were not required. On the night of 2 April, two trains were destroyed, and another three damaged. The following night two low level bomber support patrols were flown into Germany, with four Rangers sorties. Two crews, Flying Officer Dave Hewson[13] and Flight Sergeant John Hutchinson[14] (NT355), and Flying Officers John Darling[15] and Kevin Longton[16] (NT332), each destroyed trains. Flying Officers Bob Stumm[17] and Ewen Cameron[18] (NT307), in the face of determined flak, attacked a train and a signal box.

The night of 4/5 April saw a special effort required for bomber support, with 33 Mosquitos being sent off on high and low patrols, and the Bradwell Wing operating in strength, which included seven aircraft from 456. Four crews flew low level patrols, and three high level support. A total of 136 aircraft were despatched by 100 Group on this night, which was that Group's largest effort of the war.[19] Bomber Command attacked the German oil refineries at Lützkendorf and Leuna (near Leipzig), and Hamburg with 940 bombers. The results of bomber support were obvious, as the aircraft loss rate was now only 1.4%.[20] Flying Officers Hewson and Hutchinson (NT355) shot up an aircraft landing at Lüneberg, just south-east of Hamburg, but did not observe the results.

The next night for intensive bomber support patrols was 8/9 April, when 456 sent six aircraft on low level airfield patrols. The targets for the 682 bombers were again Hamburg and Lützkendorf, with a loss rate of 1.3%.[21] Again Hewson and Hutchinson (NT355) found an aircraft to shoot up on the ground when they patrolled the Jüterbog night fighter airfield, south of Berlin. An aircraft was seen coming in to land, and on finishing its landing run, put on its landing light. Hewson fired on the light, which was

11 Flt Lt G P C Rees, 424939. No 456 Sqn. Bank clerk of Lockhart, NSW; b Wagga, NSW, 3 Jan 1924.
12 Flg Off R J Letten, 418855. No 456 Sqn. Clerk of Melbourne; b Melbourne 23 Feb 1921.
13 Flt Lt D T Hewson, 419995. No 456 Sqn. Assistant engineer of Melbourne; b Melbourne 13 May 1924.
14 Plt Off J D Hutchinson, 434225. No 456 Sqn. Insurance clerk of Nedlands, WA; b Northam, WA, 12 May 1920.
15 Flt Lt J Darling, 418998. No 456 Sqn. Student of Melbourne; b Melbourne 7 Dec 1923.
16 Flt Lt K W Longton, 419192. No 456 Sqn. Bank clerk of Melbourne; b Melbourne 22 Sep 1923.
17 Flt Lt R E Stumm, 429490. No 456 Sqn. Farmer and grazier of Brookstead, Qld; b Toowoomba, Qld, 10 Apr 1924.
18 Flg Off E J Cameron, 432642. No 456 Sqn. Accountants clerk of Sydney; b Sydney 24 Aug 1919.
19 Middlebrook & Everitt, op cit, p.691.
20 ibid, p.691.
21 ibid, p.692.

immediately doused. Strikes were seen all around it, however, there was no substantiated claim.

Not many Monica contacts had been obtained by the patrols over this period, but it did prove its worth for one crew, Flight Lieutenant Col Griffin and his observer Hoppy Williams (NT311), that night:

> We were cruising home from a bomber support trip to Magdeburg in the dark early hours of April 9, 1945, when my trusty RO commanded a left, then a right, turn followed by a faintly more urgent statement that there was an aircraft closing on us dead astern and close. Who it was we didn't know; possibly a Ju 88 night fighter, perhaps another Mosquito, and more concerning, possibly a US Black Widow night fighter. The Black Widow's record of thorough friend or foe identification on black nights was lousy, so by sheer speed alone we had little trouble outrunning him, even though our concerns of dwindling fuel reserves had been something of a previous priority.
>
> Anyway, thanks to Monica, Bradwell Bay, debriefing and bed came a little sooner than our earlier ETA. Albeit, with little fuel left! As for our would-be assassin, I just wish I knew who he was.[22]

Flight Lieutenants Hoppy Williams and Col Griffin

22 Col Griffin, letter to author, of 24 Nov 1994.

The following night, 9/10 April, six aircraft supported Bomber Command on low level airfield patrols as the bombers plastered Kiel. Warrant Officer Bruce Prossor[23] and Flight Sergeant Albert Nicholas[24] (NT282) searched airfields around Wilhelmshaven, and obtained a contact but were unable to convert it to visual. However, they did attack some enemy motor transport and some lights, as Prossor recorded in his logbook:

> Patrol airfields. Hage, Ardoff, Jever. Fired at MET. No results, on Dornum-Esens r/w [railway] fired at lights. Explosion seen as passing over. On Esens-Wittmund r/w, fired at light. Explosion again as aircraft passed over. Rounds fired 240.[25]

Six patrols were flown by 456 Squadron on the night of 13 April, as the bombers returned to Kiel, the prime target this time being the U-boat yards. The following night was in support of a raid on Potsdam, when 500 Lancasters attacked the city with the prime objectives being barracks and railway facilities. This was the first time that Bomber Command four-engined "heavies" had entered the Berlin defence zone since March 1944, and was the last major raid on a German city.[26] Of the six 456 patrols, two diverted into other bases after uneventful patrols. Flying Officers John Walters[27] and Roy Woodman[28] (NT307) were forced to return single-engined, landing safely at Woodbridge, while Squadron Leader Bob Cowper and Flight Lieutenant Tom Condon (NT387) diverted into Brussels when their fuel ran short. The sorties now were at the extreme range of the fighters, with six hour patrols as the bombers targeted further east into Czechoslovakia. Intruder patrols were also flown to Norway and Denmark.

The night of 22 April saw five Rangers flown to airfields in the Munich area. Warrant Officer Walter McWhinney[29] and Flying Officer John Costigan[30] (NT332) attacked the aerodrome at Bad Aibling causing an explosion in the north-western corner of the airfield. Strikes had been seen near unidentified aircraft prior to the explosion, so a claim was lodged for one destroyed on the ground. Ground strafing was continued against

23 WOff B L Prossor, 409594. No 196 Sqn RAF, 456 Sqn. Clerk of Melbourne; b Melbourne 30 Oct 1920.

24 WOff A C Nicholas, 408469. No 456 Sqn. Asst engineer of Castle Forbes Bay, Tas; b Castle Forbes Bay 25 Oct 1923.

25 Copy from Bruce Prossor's flying logbook, quoted in letter to David Vincent from Dave Prossor, 22 Jun 1981.

26 Middlebrook & Everitt, op cit, p.695.

27 Flg Off J R Walters, 416036. No 456 Sqn. Student of Roseworthy, SA; b Adelaide 20 Sep 1921.

28 Flt Lt R A Woodman, 423275. No 456 Sqn. Clerk of Sydney; b Adamstown, NSW, 9 Aug 1921.

29 Flg Off W H McWhinney, 421614. No 456 Sqn. Packer of Forest Lodge, NSW; b Sydney 16 Mar 1921.

30 Flg Off J A R Costigan, 50325. 1 CF, No 456 Sqn. Warehouse manager of Melbourne; b Melbourne 15 Jan 1911.

Munich airfields again the next night, and on the night of 25/26 April, Flying Officers Bob Richardson and Tom Landy (NT295) claimed one unidentified aircraft damaged on the ground at Schleissheim airfield, a night fighter base on the northern outskirts of Munich. This, the night of Anzac Day 1945, was to be the Squadron's last night of operations.

The success of these long-range bomber support sorties was the reduction in numbers of bombers being shot down by *Luftwaffe* night fighters. The bomber support campaign had kept Bomber Command losses down to an acceptable level, allowing Bomber Command tactical flexibility which it would have otherwise lost through the increasing casualties inflicted on it.[31] The Mosquitos had played a vital role in the bomber offensive, without dropping bombs.

The Squadron Operations Record Book recorded: "The Armies are moving so rapidly these days that our job has virtually finished".[32]

31 Streetly, op cit, p.143.

32 No 456 Sqn Form 540, 1 May 1945.

HONOURS AND AWARDS[33]

BAR TO THE DISTINGUISHED FLYING CROSS
SQUADRON LEADER ROBERT COWPER

CITATION

This officer and Flight Lieutenant W Watson, RAF, have completed a second tour of operational duty and have consistently displayed a high degree of skill and cooperation. Their keenness to engage the enemy has always been evident and they have been responsible for the destruction of five hostile aircraft.

Flt Lt Bill Watson DFC and Sqn Ldr Bob Cowper DFC & Bar

33 London Gazette No 36931, 9 Feb 1945.

CHAPTER SIXTEEN

PEACE BREAKS OUT

Two intruder patrols to Denmark were attempted the night of 3 May 1945, but both were soon recalled with a deterioration in the weather. That was to be the last opportunity for operations, as two days later brought the news of German surrender in Holland, Denmark and northern Germany. As the Squadron waited for the news that the war was over, two aircraft were maintained at a defensive state, just in case. VE Day was Tuesday 8 May 1945, the true end of the war in Europe.

The following day was a unique occasion. Six aircraft from both of the squadrons of the Bradwell Wing, led by the station commander, Group Captain John O'Neill[1], and Flight Lieutenant Tom Condon (NT328), flew to the occupied Channel Islands as a show of force. The leader was forced to return without radio, so the formation was taken over by the CO of 456, Wing Commander Bas Howard, flying with Jack Ross (NT264). The Bradwell Wing then formed up with other Mosquitos at the mouth of the Thames, to make an aerial armada representing six squadrons and some 36 aircraft strong - "a formidable sight and noise!"[2]

The Mosquito crews had been instructed to be prepared to fire, to impress on the German commander that the war was really over, and that he should surrender to the frigate which had been sent into harbour to take over. Navigator Flight Lieutenant Bill Watson, flying this mission with Bob Cowper (NT387), recalled:

---

1 Gp Capt J A O'Neill DFC RAF, 37117. Nos 99, 58 & 238 Sqns RAF; comd 601 Sqn 1940-41; OC Flg 60 OTU 1941-42; HQ 10 Gp; comd 293 Wg 1942, 176 Sqn 1942-43, West Malling 1944, Bradwell Bay 1945. Regular air force offr; b 7 Oct 1915.

2 John Darling, letter to author, 9 May 1995 - 50 years after the event!

...evidently the German's *amour propre* demanded something more impressive (I expect he wanted a cruiser, at the least) and he had truculently threatened the frigate with his big harbour guns...it was a grand sight when the Germans laid out white crosses on the ground.[3]

Wg Cdr Bas Howard and Flt Lt Jack Ross leading the formation in NT264 RX-R.

The following day, aircraft flew to Germany, led again by Group Captain O'Neill and Tom Condon (NT387), to overfly the Ruhr and inspect the bomb damage. This was repeated the next week, on 15 May, when ten aircraft, flying a route from base over Rotterdam and Arnhem, into Germany as far as Hamm, down to Cologne, and back via Brussels. The crews observed the terrific devastation, and Ruhr overflights were repeated on 17 May, and again on 18th.

The personnel of the Squadron were now contemplating their future. The intention was initially that 456 Squadron would move to Acklington in northern England, near Newcastle, and then on to join the battle in the Far East.[4] However, in the meantime, a peacetime routine was imposed, restricting crews to 25 hours of flying training per month, with relaxed work hours which now provided for weekends. 151 Squadron moved to Predannack on 18 May, leaving the Australians as the lodger unit, and the Whitsunday long weekend allowed some of the Squadron to see the unofficial test match between Australia and England at Lords.

In the midst of this newly found freedom and uplifted spirits for the future, tragedy struck 456 Squadron the following week, on 29 May. Wing Commander Bas Howard was flying a solo sortie (NT264), to the Central Gunnery School at Catfoss, to the north near Hull, presumably to see the CO there, his old friend Group Captain Mickey Dwyer. Soon after take-off he lost an engine, and returned to land, but the undercarriage failed to lock down. On his second approach the other engine failed, and he was

3 Dr W Watson, letter to David Vincent, 17 Sep 1985.

4 No 456 Sqn Form 540, 15 & 22 May 1945.

The victory demonstration flight over the Channel Islands by the Bradwell Wing.

committed to ditching in the Blackwater River. As the aircraft hit the water, the tail section broke off and the nose plunged under the surface. Howard was evidently knocked unconscious, and drowned in the wreckage. Despite attempts to locate and rescue the CO, the efforts were to no avail. He was a much admired man, whose loss was a great shock.

Flight Lieutenant Merv Austin recorded:

> His regime was as popular as it was successful. It is difficult to analyse the secret of his hold over the unit - aircrew and groundcrew alike. He had passed through as an ordinary member and had risen by obvious merit. But it seems to have been the man himself, the kindness and strength of his character, that alone explains the unique position he held.[5]

Squadron Leader Bob Cowper took over the Squadron upon Wing Commander Bas Howard's death. The previous station commander at Church Fenton, Group Captain John Hautrey, wrote to Bob Cowper:

> I was so deeply disturbed to hear of Bas's death. As you in all probability realised, I had the very greatest admiration and affection for Bas, an admiration and affection which was also extended to his Squadron which seemed to me [had] taken to itself all those high ideals and that quiet efficiency which Basil always stood for.[6]

5 Flt Lt Austin, op cit, Part VI, p.14.

6 Gp Capt J Hautrey, Air Component SHAEF France, to Sqn Ldr Bob Cowper, 5 Jun 1945. Courtesy of Bob Cowper.

Wing Commander Bas Howard DFC

This was not the only loss to the Squadron during its final days. LAC Frank Gregson[7], a Fitter IIE, who had already been recognised with a mentioned in dispatches for his service in the Middle East, was to die of illness. He had contracted tuberculous meningitis, and passed away on 8 June, in hospital at Haywards Heath, in Sussex.

No 456 Squadron was disbanded a week later, on 15 June 1945. It had been a night fighter squadron, the only one ever in the existence of the RAAF, that was the only Australian unit of the war to have extensive experience with AI radar. Furthermore, the Squadron had flown varied missions, day and night, for four years, from 1941 until the end of the war. These operations had been conducted from the quiet sector in Wales, to the front-line of Normandy, as far south as the Spanish border, north to Norway, and into the heart of Germany, to Berlin. The Australians had flown the hopelessly inadequate Defiant, then the first true night fighter, the Beaufighter, and that wonderful multi-role aircraft, the Mosquito, which they had proved could truly deliver a sting. One night fighter pilot summed up the difficulty of their mission:

> ...a pilot had to be able to fly the aeroplane on instruments almost as violently as a day fighter in sunshine. Add low cloud, rain, or perhaps fog and the perennial fear of icing up, and then without knowing from quite where he started, find the way a hundred miles in absolute darkness and

7 Cpl F Gregson, 11072. 1 AD, 1 ED, 205 Gp RAF, 462 & 456 Sqns, 6456 SE. Engineer of Melbourne; b Melbourne 17 May 1917. Died of illness 8 Jun 1945.

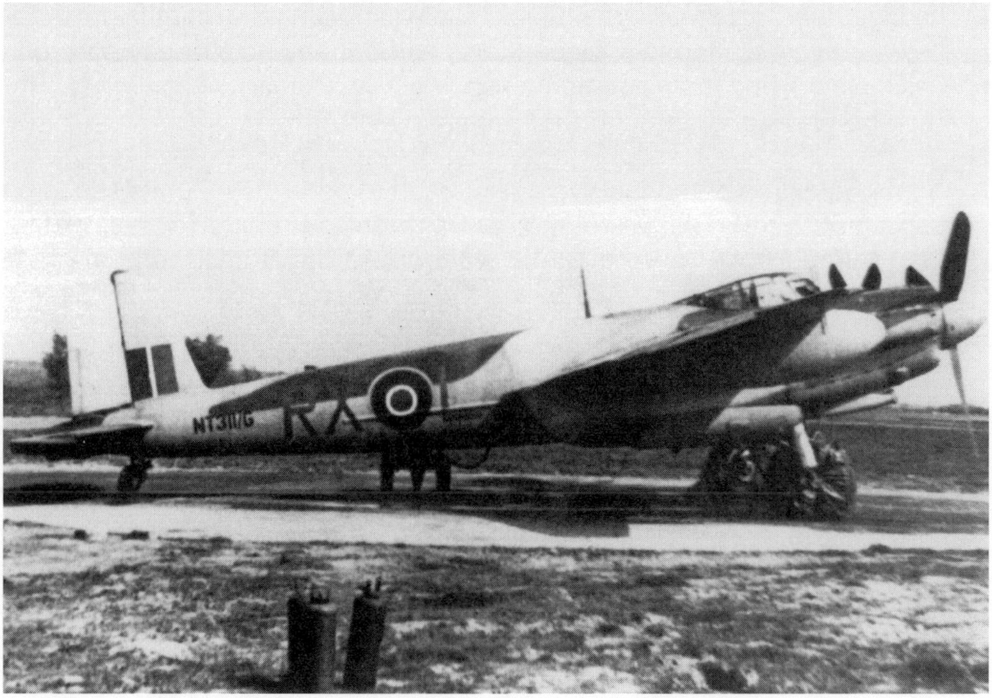

Above: Mosquito NT311 RX-L. Visible is the G suffix to the serial number (NT311/G), because of the sensitivity of the radar, indicating the necessity for the aircraft to be guarded when on the ground.
Below: Mosquito MM687 RX-Z served with 456 Squadron from December 1944 until April 1945. Postwar, this aircraft was transferred to the Belgian Air Force as MB-21.

> bad weather to a small airfield which only shows its few faint lights when it is sure you are friendly. Once there, without any help of any sort, only basic blind flying instruments (which you always fear will let you down), feel the way down through the murk and land.[8]

Some Squadron personnel were dispersed to other RAAF units serving in Europe, but in the main were sent to Brighton to prepare for repatriation. Most returned home on the *SS Stratheden*, sailing from Liverpool on 19 October, and arriving in Australia - "our most glorious sight, Sydney Harbour" - in late November.[9]

Although No 456 Squadron had devised the now familiar RAAF kangaroo roundel, it was unfortunate that the Squadron never had an official badge approved. If it had, 456 could not have gone past the motto of one of its sister night fighter squadrons, No 604 Squadron RAF, whose motto of "*Si vis pacem para bellum*" was most appropriate - "If you want peace, prepare for war". The crews of 456 Squadron had prepared during their quiet periods, and when faced with the demands asked of them, they had been ready. In four years of operations, No 456 Squadron had flown over 6,200 hours on operations. It had been a proud record, acquitted with distinction.

8 Meadows, Part 5, in Aeroplane Monthly, Feb 1993, p.23.

9 Len Stamp, letter to author, 11 Dec 1994.

EPILOGUE

By GROUP CAPTAIN DAVE SCHUBERT

In 1986, as President of the Mess Committee at Point Cook, I had the honour of meeting Wing Commander Gordon Olive DFC, when officiating at a reunion dining-in night for No 456 Squadron RAAF. The dinner, a display by the RAAF Museum about 456 Squadron, and the presence of the son of a former RAF CO representing his father and the RAF interest, boded well for a good evening of accurate history telling. In particular, they looked to their former CO with pride and respect.

As Wing Commander Olive rose to speak after dinner, most of the Squadron members strained to see the CO who had been racked by illness which cut his tour short in March 1942. Olive spoke of the hard times and the great effort they had all made to overcome incredible difficulties and the nature of war. He spoke of the improving morale and the later achievements of 456 when they re-equipped with Mosquito night interceptors. He mentioned that a later RAF CO of the Squadron [Wing Commander Dwyer] - there being no suitably qualified Australians available - was so impressed with the elan of the Squadron that he elected to wear an RAAF brevet while commanding 456 Squadron.

They took in his every word out of awe and respect for the man. No one moved or looked away. It was an inspiration to see the feeling amongst the gathering and learn about loyalty and leadership. I regret to say that Wing Commander Olive DFC passed away the following year.

APPENDIX 1

HONOUR ROLL

NO 456 SQUADRON RAAF

AUSTRALIANS

LAC F W Bidgood	Died of Illness	4 October 1941
Flight Lieutenant A R Parker	Killed in Flying Accident	5 January 1942
Flight Sergeant R N B Scott	Killed in Flying Accident	8 October 1942
Sergeant G A C Wood	Killed in Flying Accident	8 October 1942
Flying Officer D L McGregor	Killed in Flying Accident	1 November 1942
Sergeant L M McCormack	Killed in Flying Accident	1 November 1942
Sergeant W R G Anderson	Killed in Flying Accident	30 January 1943
Sergeant L Gill	Killed in Flying Accident	30 January 1943
Flight Sergeant C Stovell	Died of Illness	26 February 1943
Flying Officer R B Hitchcock	Killed in Flying Accident	23 May 1943
Pilot Officer R C Lowther	Killed in Action	2 June 1943
Warrant Officer V P Ratcliffe	Killed in Action	2 June 1943
Flight Lieutenant E H Griffith	Killed in Flying Accident	29 September 1943
Corporal W H Blakeley	Killed in Flying Accident	29 September 1943
Flying Officer L R Parnell	Killed in Action	4 December 1943
Pilot Officer J L May	Killed in Action	4 December 1943
Flying Officer R M J Pahlow	Killed in Action	29 April 1944
Flying Officer F M Silva	Killed in Action	29 April 1944
Flying Officer E C Radford	Killed in Action	12 July 1944
Pilot Officer W E Atkinson	Killed in Action	12 July 1944
Warrant Officer J L Mulhall	Killed in Action	24 Nov 1944
Wing Commander B Howard	Killed in Flying Accident	29 May 1945
Corporal F Gregson	Died of Illness	8 June 1945

BRITISH

Sergeant A F Brookes	Killed in Flying Accident	21 July 1941
LAC H R Roberts	Killed in Flying Accident	23 May 1943
Flying Officer J D Jones	Killed in Action	24 November 1944

AMERICAN

2nd Lieutenant F W Fraser	Killed in Flying Accident	5 January 1942
Lieutenant E M Woodward	Killed in Action	7 November 1944
Ensign W G Madden	Killed in Action	7 November 1944

Fred Stevens at the 456 Squadron Roll of Honour, Australian War Memorial, Canberra, 1994

APPENDIX 2

No 456 SQUADRON

SQUADRON CHRONOLOGY

30.6.41	Sqn Ldr C G C Olive DFC	Valley	Defiant I
26.9.41			Defiant/Beaufighter II
8.10.41	WgCdr C G C Olive DFC		
31.10.41			Beaufighter II
1.3.42	Sqn Ldr J S Hamilton (Temp)		
2.4.42	WgCdr E C Wolfe DFC		
7.7.42			Beaufighter II/VI
20.8.42			Beaufighter VI
31.12.42			BeauVI/Mosquito II
18.1.43			Mosquito II
1.2.43	WgCdr M H Dwyer		
30.3.43		Middle Wallop	
12.6.43	WgCdr G Howden DFC		
16.7.43			Mosquito II/VI
17.8.43		Colerne	
10.10.43			Mosquito II
17.11.43		Fairwood Common	
14.12.43	Wg Cdr K M Hampshire DSO		
29.1.44			Mosquito II/XVII
1.3.44		Ford	Mosquito XVII
12.5.44	Wg Cdr K M Hampshire DSO DFC		
10.11.44	WgCdr B Howard		
7.12.44			Mosquito XVII/XXX
31.12.44		Church Fenton	
9.2.45	Wg Cdr B Howard DFC		
15.2.45			Mosquito XXX
17.3.45		Bradwell Bay	
29.5.45	Sqn Ldr R B Cowper DFC*		
15.6.45	SQUADRON DISBANDED		

APPENDIX 3

AWARDS

No 456 SQUADRON RAAF

BAR TO DISTINGUISHED SERVICE ORDER (DSO*)
Wg Cdr K McD Hampshire DSO DFC

OFFICER OF THE ORDER OF THE BRITISH EMPIRE (OBE)
Sqn Ldr S M Green

BAR TO DISTINGUISHED FLYING CROSS (DFC*)
Sqn Ldr R B Cowper DFC

DISTINGUISHED FLYING CROSS (DFC)
Wg Cdr K McD Hampshire DSO
Wg Cdr B Howard
Sqn Ldr G M Robinson RAF
Flt Lt F S Stevens
Flt Lt T Condon
Flt Lt G Panitz
Flt Lt J R Ross
Flg Off I W Sanderson
Flt Lt W Watson RAF
Flg Off R S Williams

BRITISH EMPIRE MEDAL (BEM)
Flt Sgt E Barber
Flt Sgt J H Goodall
Sgt J J Plunkett

MENTIONED IN DISPATCHES (MID)

Flt Lt J Darling
Flt Lt S M Green
Flt Lt J E Pike
Flt Lt R G Pratt
Flt Lt J R Ross
Flt Lt S D P Smith
Flt Sgt J H Goodall
Flt Sgt F S Stevens
Flt Sgt L M Sutherland

APPENDIX 4

BOULTON PAUL DEFIANT AIRCRAFT OF No 456 SQUADRON

The Defiant was the standard night fighter in Britain during 1941, having been relegated from day operations because of its vulnerability. The first Defiants were received by No 456 Squadron at Valley, in Anglesey, Wales, at the beginning of July 1941, and the type was operated by the Squadron for four months, with the last operation on 20 October 1941.

BOULTON PAUL Defiant Mark I (Jul-Oct 1941)

Serial	Date On	Code	Date Off	Fate
N1558	4.7.41		24.10.41	to 60 OTU, converted to TT mk III
N1569	4.7.41		26.7.41	damaged on landing Valley (Wild), to 43 Group[1] for repair
N1647	4.7.41		21.10.41	to 60 OTU, written-off 25.1.42 on 153 Sqn
N1690	2.7.41		26.9.41	damaged on landing Valley 17.9.41 (West), 43 Gp for repair
N1706	7.9.41		24.10.41	to 60 OTU, then 153 Sqn
N3313	10.9.41		15.11.41	to 43 Gp depot (ex PS-P of 264 Sqn)
N3367	4.7.41		14.10.41	to 256 Sqn, convtd to TT.mkIII 30.11.43
N3370	10.9.41		4.11.41	to 125 Sqn, convtd to TT.mkIII 25.1.44
N3458	29.8.41		5.12.41	damaged on landing Valley 3.9.41 (Kirkman), repaired 43 Gp, to 256 Sqn
N3477	17.9.41	PZ-V	31.10.41	to 125 Sqn, then 286 Sqn

1 RAF Maintenance Command, like the other Commands, was divided into Groups. No 40 Gp provided all non-explosive equipment, and maintained Aircraft Equipment Depots. No 41 Gp received aircraft from factories, incorporated service modifications, and forwarded them to the squadrons. No 42 Gp issued bombs and explosives. No 43 Gp was responsible for all repairs that were beyond the capability of the squadrons.

Serial	Date On	Code	Date Off	Fate
N3496	2.7.41		.41	to servicing Reid & Sigrist[2], then 24 MU 2.1.42, converted to TT.mkIII 12.4.44
N3497	2.7.41		16.9.41	damaged on landing Valley 6.9.41 (Pargeter/Fell), to repair at Reid & Sigrist, convtd TT.mkIII 12.1.44
T3933	4.7.41		21.7.41	crashed Llanerchymedd, Anglesey, (Brookes killed), written-off
T3940	4.7.41	PZ-V	30.10.41	crashed on landing Valley 27.7.41 (Woodill), repaired on 456, to 125 Sqn
T3950	2.7.41		27.8.41	damaged on landing Valley 20.8.41 (Gloster), to repair at Reid & Sigrist
T4003	9.9.41		21.10.41	to 60 OTU, written-off 31.10.41
T4007	4.7.41		23.8.41	damaged on landing Valley 15.8.41 (Dodson), to repair at Reid & Sigrist
T4065	18.9.41		21.10.41	to 60 OTU, damaged 26.2.42
T4066	29.7.41		25.10.41	to 256 Sqn (coded JT-P)
V1116	16.9.41		25.10.41	to 256 Sqn (JT-H), written-off 7.2.42
V1124	5.8.41		25.10.41	to 256 Sqn, damaged 23.11.41, converted to TT.mkIII to Royal Navy 19.1.45
V1175	2.7.41		24.10.41	to 60 OTU, then 153 Sqn and written-off 13.11.41
V1177	2.7.41		4.9.41	forced landed and hit trees at Squires Gate, Blackpool (Cleary), written-off
AA282	4.7.41		25.10.41	to 256 Sqn (coded JT-Q), converted to TT.mkIII 21.9.43
AA283	2.7.41		16.11.41	to servicing at Reid & Sigrist, then 2AFU, written-off 29.9.42
AA284	2.7.41		24.10.41	to 60 OTU, then 277 Sqn, written-off 25.10.42
AA285	2.7.41		20.10.41	to 60 OTU, damaged 4.1.42
AA286	17.9.41		25.10.41	to 256 Sqn (coded JT-K), converted to TT.mk III 8.12.43

(Note: There is no official record of aircraft N3430 or T4111, sometimes quoted as 456 Squadron aircraft, of ever serving on the unit.)

Most of 456 Sqn's remaining Defiants were passed to No 256 Sqn (based at Squires Gate, Blackpool) or to No 60 Operational Training Unit (60 OTU), at RAF East Fortune, East Lothian, at the end of operations in October 1941. Many of the aircraft that survived were converted to the target towing Mark III variant over 1943-44.

Only one Defiant now survives. Aircraft N1671 is on display at the RAF Museum, Hendon, in London, marked as EW-D of No 307 (Polish) Sqn, when based at Kirton-in-Lindsey, Lincolnshire.

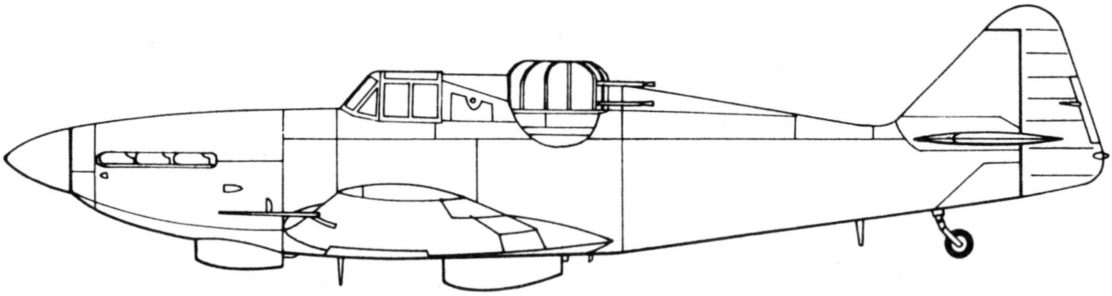

2 Much of Defiant servicing was contracted to Reid & Sigrist Ltd, at Desford Aerodrome, Leicester.

APPENDIX 5

TRAINING AIRCRAFT OF No 456 SQUADRON

The changeover from the Defiant to the twin-engined Beaufighter necessitated twin-engined training for the Squadron's pilots. In addition several aircraft were held on the unit as "hacks", to provide a taxi service and continuity flying.

BRISTOL 142M Blenheim Mark I

During the period of Defiant operations, No 456 Sqn received a Bristol Blenheim (L4907, powered by Mercury VIII radials), from 54 OTU, for twin-engined training for the conversion of pilots to the Beaufighter. This aicraft was subsequently joined during 1942 by a second Blenheim from 68 Squadron (L1371), and by a third, L1170 from 51 OTU (both powered by Mercury XVs), so that from June until September, 456 Sqn had three of the type on strength.

Serial	date on 456	Date off 456	Fate
L1170	21.5.42	5.10.42	to 409 Sqn Coleby Grange
L1371	.4.42	13.5.42	loan to 96 Sqn wrexham
	.6.42	.9.42	to 488 Sqn Church Fenton, struck off charge 1.7.43
L4907	28.8.41	11.1.43	coded '20' from 54 OTU, forced landed Valley 2.4.42 (Kerr), damaged on ground 10.2.43
	18.4.43	23.6.43	to 55 OTU, struck off charge 9.7.43

AIRSPEED AS.10 Oxford Mark II

P8606	28.6.42	14.6.45	damaged when forced landed in field nr Petersfield, Hampshire (Houston)

MILES M.14A Magister Mark I

P2464	.10.41	23.5.43	crashed Middle Wallop (Hitchcock/Roberts killed)
T9848	.6.43	15.5.43	damaged Fairwood Common 14.2.44 (Houston), to 151 Sqn Bradwell Bay, sold 13.5.46, became G-AHYM 18.3.47

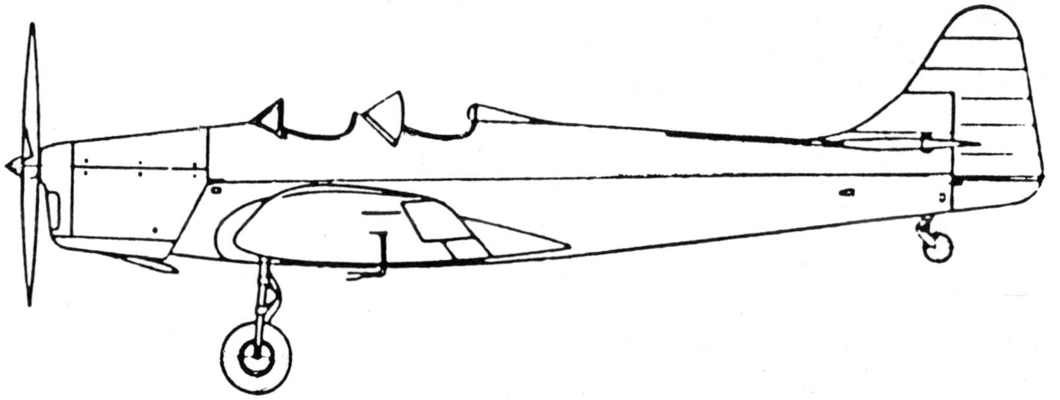

Miles Magister Mark I

APPENDIX 6

BRISTOL BEAUFIGHTER AIRCRAFT OF No 456 SQUADRON

The Mk II was the was the only variant of the Beaufighter powered by the ubiquitous Rolls-Royce Merlin (two 1280hp Merlin XX liquid-cooled piston engines) to enter operational service. The use of the Merlin in the Beaufighter was only intended as a stopgap until the Bristol Hercules radial air-cooled engine became available. Three prototypes and 447 production models of the IIF variant were manufactured by the Bristol plant at Filton. The first production Beaufighter IIF (R2270), the "F" suffix indicating its fighter role, flew on 22 March 1941.

The first Beaufighter IIF for No 456 Squadron (R2474) arrived at Valley on 26 September 1941. At this stage, the standard night fighter colour scheme was RDM2 Soot Black. The squadron code for 456 was changed from PZ, as this was also being used by No 53 Squadron RAF, initially to SA, then in early 1942 to RX. (The allocation of SA may have been in error, as No 486 (NZ) Sqn then adopted this code on forming with Hurricanes in March 1942.) Aircraft assigned to 456's 'A' Flight carried individual letters A to L, those with 'B' Flight were M to Z. Conversions for crews on the Squadron commenced at the beginning of October 1941, and the first operational sortie was flown on 29 November 1941.

BRISTOL 156 Beaufighter Mark IIF (Sep 41 - Jul 42)

Serial	Date On	Code	Date Off	Fate
R2325	5.5.42	RX-K	18.1.43	to 54 OTU, damaged 17.2.43
R2334	5.10.41		18.11.41	crashed landing Valley (Kirkman), aircraft cartwheeled, struck off charge 6.12.41
R2341	4.10.41		21.8.42	damaged landing Valley 25.2.42 (Dey), damaged on transfer to 60 OTU High Ercall
R2343	12.10.41		5.4.43	damaged ldg Valley 5.5.42 (Scott), repaired on 456, to 54 OTU Charter Hall
R2344	12.10.41	RX-L	4.9.42	to 54 OTU, damaged 2.1.43
R2371	12.10.41	RX-R	19.9.42	to 132 OTU, damaged 14.5.43

Serial	Date On	Code	Date Off	Fate
R2372	12.10.41		23.11.41	damaged landing Valley 20.11.41 (Cleary), to Bristol for repair
R2381	21.10.41		3.5.42	damaged landing Speke 25.2.42 (Ward), damaged on take-off 14.3.42 (Kerr), crashed on take-off Valley (Hough) and written-off
R2382	29.11.41	RX-Z	31.7.42	first 456 aircraft with kangaroo marking, to 54 OTU, damaged 4.6.43
R2394	26.3.42		18.11.42	damaged landing Valley 14.5.42 (Panitz), to 27 MU then 63 OTU
R2395	9.1.42		20.8.42	to 132 OTU East Fortune
R2474	26.9.41		5.1.42	crashed in sea nr Valley (Parker/Fraser killed), written-off
R2476	10.10.41		5.1.42	damaged on forced landing at Trewan Sands approaching Valley (Hedge/Hancock), reassessed as write-off and struck off charge 14.5.42
T3012	27.9.41	RX-K	20.3.42	crashed on take-off Valley (Spring) and written-off
T3014	27.9.41		26.6.42	shot down by Ju 88, ditched, crew safe (Dey/ Mitchell)
T3017	30.9.41	RX-B	21.7.42	to 54 OTU, damaged 28.2.43
T3021	30.9.41	RX-J	8.5.43	damaged landing 5.5.42 (Dey), damaged taxying Valley 4.8.42 (Timbury), to 54 OTU
T3022	27.9.41		10.8.42	to 60 OTU, written-off 27.4.43
T3024	7.10.41		8.12.41	crashed landing Valley 15.11.41 (Kirkman), to Bristol for repair
T3026	7.10.41		27.4.42	wheels-up landing Valley 31.1.42 (Hedge), repaired to 307 Sqn Exeter
T3027	7.10.41	RX-H	19.9.42	to 132 OTU
T3028	7.10.41	RX-V	20.3.43	damaged landing Valley 5.1.42 (Cleary), to 54 OTU
T3036	28.1.42		10.8.42	damaged landing Valley 2.4.42 (Panitz/Cloke), to 132 OTU
T3042	25.11.41		20.8.42	to 60 OTU, damaged 16.3.43
T3047	25.11.41		16.2.42	crashed landing Valley 28.1.42 (Hughes), repaired at Burnley Aircraft, to 63 OTU
T3145	13.6.42	RX-F	31.7.42	to 132 OTU, 54 OTU, written-off 25.9.43
T3151	9.1.42		17.8.42	damaged landing Valley 16.2.42 (Cleary), to 132 OTU, damaged 8.5.43
T3181	9.5.42		21.7.42	to 54 OTU, damaged 3.8.42
T3368	3.6.42		4.9.42	to 410 Sqn Scorton
T3370	29.3.42	RX-Z	24.10.42	damaged landing Valley 14.5.42 (Dottridge) and 4.6.42 (Dottridge), to 54 OTU, written off 31.7.43
T3371	16.5.42	RX-D	21.8.42	to 54 OTU, damaged 16.11.42
T3379	26.3.42		12.7.42	damaged landing Valley 30.4.42 (Ward), to 54 OTU
T3383	26.3.42	RX-A	17.9.43	damaged landing Valley 28.10.42 (Timbury/ Skippen), repaired on 456, to 63 OTU
V8133	8.5.42	RX-E	17.8.42	damaged landing Valley 17.7.42 (Ward), to 54 OTU

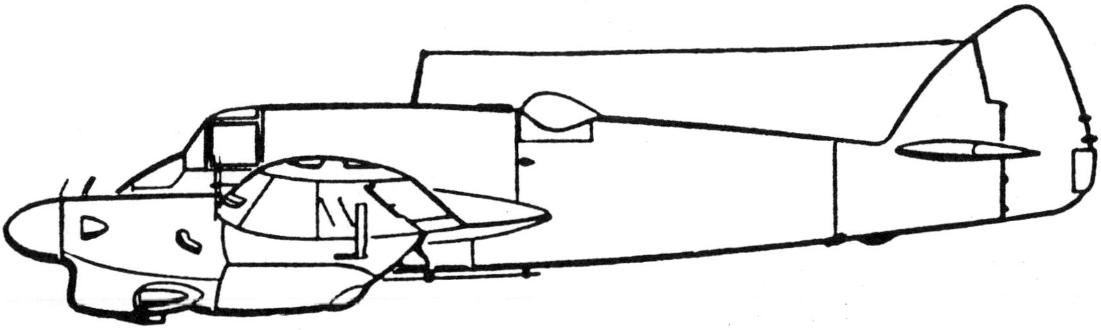

Beaufighter Mark IIF

All 456's Beaufighter IIs retained the overall black scheme. At least one aircraft, R2382 RX-Z, had a red kangaroo painted on the nose. This was not very conspicuous, and in 1943 the Squadron was to adopt the kangroo roundel as an unofficial marking. The CO's aircraft, T3383 RX-A, carried a Wing Commander's pennant on the nose. The last 456 Beaufighter II operation was flown from Valley on 20 August 1942, with some aircraft remaining on 456 for several months as trainers, until being tranferred to OTUs.

BRISTOL 156 Beaufighter Mark VIF (Jul 42 - Jan 43)

No 456 Squadron's first Beaufighter VI (X8251), powered by Bristol Hercules 6 radials, was received on 7 July 1942, and within ten days, a dozen had been delivered. The first operation was flown by 456 on 25 July. The Squadron was advised on 15 August 1942 of a new camouflage scheme for its night fighting Beaufighters. The overall black was replaced by the day fighter colours on upper surfaces of dark green and ocean grey. Undersurfaces remained matt black, and the squadron code letters dull red. The last aircraft were not stripped and repainted until December 1942.

V8440	24.10.42		25.1.43	to 96 Sqn Honiley, damaged 12.3.43
V8441	11.10.42		7.1.43	to 96 Sqn, to instr a/f 3.10.45 as 5696M
V8444	13.10.42	RX-W	1.11.42	missing Irish Sea (McGregor/McCormack)
V8455	16.10.42	RX-G	23.1.43	to 96 Sqn, damaged 10.8.43
V8467	11.10.42	RX-H	6.5.43	damaged on take-off Valley 26.10.42 (Richardson), repaired on 456, to 406 Sqn
V8495	11.10.42	RX-S	9.1.43	forced landed Valley 30.11.42 (Pargeter/ Fell), to 96 Sqn, damaged 21.10.43
V8520	19.11.42		7.1.43	to 96 Sqn (coded ZJ-B), to instr a/f in 1945 as 5619M
X7895	13.7.42	RX-V	13.11.43	forced landed Omagh, Nthn Ireland 19.11.42 (Ratcliffe/Abbey), damaged 13.11.43 to repair
X7896	15.7.42		20.10.43	damaged landing Valley 6.10.42 (Panitz/Williams), repaired on 456, to 19 MU
X8026	7.11.42	RX-L	28.1.43	to 256 Sqn Woodvale, written-off 10.3.43
X8100	19.7.42	RX-E	5.6.43	damaged 14.1.43, repaired on 456, to 125 Sqn, then 96 Sqn
X8138	18.7.42	RX-Z	5.6.43	"Bunyip", damaged by fire on ground 9.11.42, to 125 Sqn Exeter
X8190	16.7.42	RX-K	8.10.42	crashed Valley (Scott/Wood killed)
X8194	4.8.42	RX-T	10.1.43	crashed nr Amlwch, northern Anglesey (Samson), written-off
X8195	26.7.42		30.7.43	damaged landing Valley 20.9.42 (Pargeter/Ashcroft), repaired on 456, to 141 Sqn Wittering
X8199	17.8.42	RX-J	24.3.43	forced landing on approach Valley 25.2.43 Richards/Houston), to repair at 19 MU
X8200	18.7.42		28.10.43	damaged take-off Valley 2.10.42 (Pargeter/Fell), repaired 456, damaged 21.10.43, to repair at 32 MU
X8205	26.7.42	RX-H	14.9.42	forced landing nr Valley (Spring/Thomson), caught fire and written-off
X8206	18.7.42	RX-G	15.10.42	crashed on approach Valley (Smith/Stovell)
X8251	10.7.42		28.1.43	forced landed .12.42 (Wolfe), to 256 Sqn, damaged 16.6.43
X8261	18.7.42	RX-F	18.4.43	to 406 Sqn Valley, written-off 7.6.43
X8262	18.7.42	RX-Y	16.6.43	damaged 16.6.43, to repair by Southern A/c
X8263	14.7.42	RX-D	7.3.43	to 488 Sqn Ayr, written-off 2.4.43
X8265	13.7.42	RX-B	18.4.43	to 406 Sqn Valley
EL188	11.10.42	RX-R	7.1.43	to 96 Sqn (coded ZJ-H), then 406 Sqn
EL214	13.10.42	RX-K	30.1.43	crashed in Irish Sea (Anderson/Gill killed)

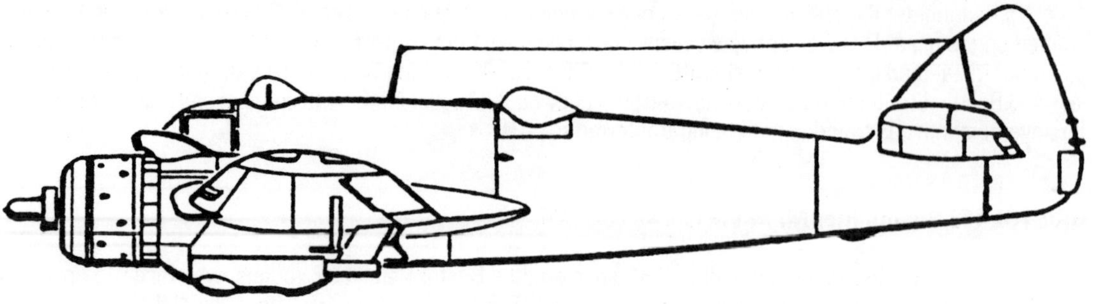

Beaufighter Mark VIF

One aircraft, Fred Steven's X8138 RX-Z, was painted as the "Bunyip", and it was intended that all 'B' Flight aircraft be christened with an Australian animal name beginning with "B". These plans ran out of steam, presumably because of a lack of "B" animals! Other aircraft did carry individual markings, such as "Gate" Gatenby's Australian thumbs-up marking "Up 26", on the nose of X8026 RX-L.

With the arrival of the first Mosquitos in January 1943, 456 began disposing of its Beaufighters. Some of the remaining Beaus were delivered from Valley to No 96 Sqn, based at Honiley, Warwickshire, and those remaining at Valley were transferred to No 406 Sqn RCAF. The last 456 Squadron Beaufighter operational sortie was flown on 18 January 1943.

APPENDIX 7

DE HAVILLAND MOSQUITO AIRCRAFT OF No 456 SQUADRON

No 456 Squadron was advised of re-equipment with the Mosquito Mark II on 2 December 1942. The day fighter variant, the Mosquito F.II, had been modified with the addition of a Mark IV AI radar, and this night fighter version was the NF.II.

DE HAVILLAND DH-98 Mosquito NF.II (Jan 43 - Mar 44)

The first three Mosquito NF.II aircraft for No 456 Sqn arrived at RAF Valley on New Years Eve 1942. A total of 494 Mosquito Mark IIs, powered by Merlin series 20 engines, were built.

In May 1943, 456 Sqn moved to Middle Wallop, Hampshire, to swap with a Canadian night fighter unit, No 406 Sqn. Three months later, in August, No 456 was transferred again, to swap places with No 151 Sqn at the 10 Group Fighter Command Sector Station, at RAF Colerne, in Wiltshire.

Serial	Date On	Code	Date Off	Fate
DD605	6.9.43		4.11.43	to 141 Sqn Wittering, 169 Sqn, damaged 5.1.44
DD610	24.8.43	RX-G	28.2.44	to 13 MU for maintenance
DD614	1.5.43		1.8.43	loaned to 140 Sqn crash landed 9.8.43 Eversley Cross, Surrey (Gilbert), written-off
DD617	22.9.43		25.11.43	for repairs, to 25 Sqn, became instr a/f 1.4.45 as 4360M
DD620	1.5.43	RX-O	27.2.44	to 13 MU for maintenance, to 51 OTU Cranfield
DD622	.1.44	RX-M	12.2.44	to 218 MU for maintenance, to 239 Sqn

Serial	Date On	Code	Date Off	Fate
DD626	16.7.43		4.9.43	to 140 Sqn Hartfordbridge
DD682	28.1.43		16.2.43	damaged 29.1.43, to De Havillands for repair
DD719	20.12.43	RX-U	27.2.44	to 13 MU, then to 51 OTU
DD726	17.4.43	RX-M	6.10.43	to 140 Sqn
DD734	20.12.43		30.12.43	to 157 Sqn Predannack, 13 OTU, written-off 7.11.44
DD739	12.4.43	RX-X,-C	4.12.43	failed to return from "Mahmoud" operation to Kassel (May/Parnell)
DD751	25.1.43	RX-W	4.3.44	to 13 MU, then to 51 OTU
DD786	13.12.43		16.3.44	damaged landing Cranfield, Bucks, 28.2.44 (Mzourek), to 13 OTU, written-off 14.4.44
DD789	14.1.43	RX-D	9.6.43	damaged landing Middle Wallop 30.5.43 (Smith), to repair
DZ247	31.12.42		14.4.43	damaged landing Middle Wallop 3.4.43 (Houston), to repair
DZ251	13.12.43	RX-R	25.1.44	to 157 Sqn, then to 60 OTU
DZ265	23.9.43	RX-T	10.2.44	to 218 MU, then to 239 Sqn
DZ267	30.1.43	RX-Z	4.11.43	to 141 Sqn (coded TW-Y), damaged 6.1.44, then to 169 Sqn
DZ268	.1.43	RX-B	4.2.44	to 157 Sqn, then to 60 OTU
DZ269	9.1.43	RX-U	1.10.43	to 60 OTU High Ercall, damaged 3.12.44
DZ293	9.1.43	RX-A	30.4.44	to Ftr Interception Unit, Ford, then to 54 OTU
DZ297	31.12.42	RX-R	4.11.43	forced landed 20.2.43 (Stevens), repaired on 456, to 141 Sqn, damaged 6.1.44
DZ299	5.6.43	RX-D	23.7.43	to FIU, Ford, damaged 6.9.43, to 51 OTU
DZ307	29.1.43	RX-Y	22.4.44	damaged landing Colerne 16.11.43 (Wilkie), to 157 Sqn Valley
DZ308	24.1.43	RX-V	2.6.43	failed to return from "Ranger" op France (Ratcliffe/Lowther)
DZ310	22.1.43	RX-T	27.7.43	damaged landing Warmwell, Hants, 17.5.43 (Palmer), DeHav for repair, to 169 Sqn (coded VI-B), then to 141 Sqn (TW-Z)
DZ658	14.1.43		25.1.43	crashed on approach Valley (Hitcock/Parnell), written-off
DZ681	1.2.43	RX-L,-V	15.1.44	overshot runway Fairwood Common (Radford/ Atkinson), written-off
DZ682	23.1.43	RX-K	25.1.44	to 157 Sqn, damaged 20.2.44
DZ690	16.7.43		29.9.43	crashed after engine fell off, Colerne (Griffith/Blakeley killed)
HJ650	5.4.43	RX-S	14.2.44	to 157 Sqn, written-off 22.11.44
HJ701	4.5.43	RX-C	30.5.43	engine failed and forced landed Crediton, Devon (Dwyer/Shanks injured), written-off
HJ702	4.5.43	RX-X	23.7.43	to FIU, Ford, then to 13 OTU
HJ916	21.1.43	RX-J	26.4.43	damaged landing Middle Wallop 4.4.43 (Pratt), to 71 MU for repair
HJ925	29.4.43	RX-P	29.2.44	damaged landing Colerne 20.9.43 (Thomas), repaired on 456, to 13 MU
HJ934	12.12.43	RX-W	10.2.44	to 218 MU, then to 239 Sqn
HJ938	12.12.43	RX-H	29.2.44	to 13 MU, then to 51 OTU
HJ939	3.1.43	RX-E	15.9.43	to 60 OTU, then to 54 OTU
HJ941	22.1.43	RX-G	13.11.43	to 141 Sqn, missing on ops 28.6.44

Serial	Date On	Code	Date Off	Fate
HJ942	12.4.43	RX-H,-J	13.1.44	to Fairwood Common Stn, to 13 OTU
HJ943	21.1.43	RX-F	11.11.43	to 141 Sqn, missing on ops 25.2.44

Colour scheme for 456 Sqn Mosquitos was Medium Sea Grey overall, with a disruptive pattern of Dark Green on the uppersurfaces, and Dull Red code letters. The Squadron remained at Colerne until being replaced by No 151 Sqn in November 1943. 456 then moved to its new base at Fairwood Common, near Swansea in Glamorgan, where it replaced No 264 Sqn.

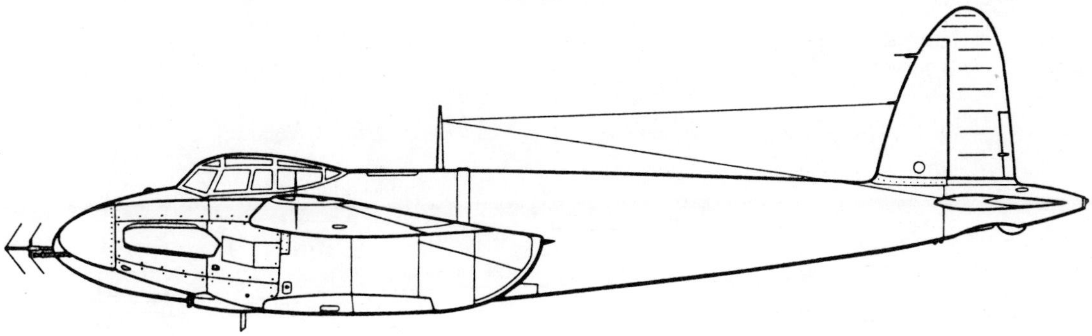

Mosquito NF.II

DH-98 Mosquito T.III (Jan 43 - Jul 44)

The T.III was the dual-control version of the Mosquito F.II, with all armament removed and used for pilot training, and powered by Merlin 23s. A total of 358 T.IIIs were built.

HJ852	22.1.43	RX-M	25.5.43	forced landed Valley 23.2.43 (Maloney), to 60 OTU
HJ899	7.4.44	RX-V	7.7.44	to 51 OTU Cranfield, to 13 OTU

Most Mosquito T.IIIs were used by the training units. Initially there were no Mosquito OTUs, the first unit being the Mosquito Conversion Unit at RAF Horsham St Faith, Norfolk, formed in August 1942. Eventually, the following units were equipped with Mosquitos:

No 8 OTU, RAF Fraserburgh, Dyce
No 13 OTU, RAF Harwell, Berkshire
No 51 OTU, RAF Cranfield, Bedfordshire
No 54 OTU, RAF Charter Hall, Berwickshire
No 60 OTU, RAF High Ercall, Shropshire
No 132 OTU, RAF East Fortune, East Lothian.

DH-98 Mosquito FB.VI (Jul 43 -Oct 43)

456 Squadron had been advised on 28 February 1943 that a new role would be intruder operations over France as part of "Operation Ranger". For these Ranger missions, a flight of an additional six aircraft were to be allotted to the unit. Initially Ranger sorties were flown in the NF.II aircraft, but finally six FB.VI were delivered in July 1943, from No 151 Sqn at Colerne. Although similar to the NF.II, the FB.VI had no AI radar, but were better equipped for the role with long-range fuel tanks and bombing equipment. The aircraft were flown extensively on detachment from Predannack, on Ranger and Instep patrols, the latter to provide cover for Coastal Command aircraft in the Bay of Biscay.

Serial	Date On	Code	Date Off	Fate
HJ814	20.7.43	RX-D	10.10.43	to 60 OTU High Ercall
HJ815	20.7.43	RX-N	10.10.43	to 60 OTU
HJ816	20.7.43	RX-X	8.10.43	dam by flak 7.8.43 France (Samson/Abbey), to 60 OTU, written-off 10.9.44
HJ817	16.7.43	RX-E	8.10.43	to 60 OTU and written-off 14.10.43
HJ818	16.7.43	RX-L	8.10.43	dam by flak 27.7.43 France (Gatenby/Fraser), to 60 OTU and written-off 1.5.44
HJ819	16.7.43	RX-V	8.10.43	to 60 OTU

The FB.VI was powered by the Merlin 21 or 23, and was operated by 456 Squadron until October 1943, when all six aircraft were delivered to No 60 OTU, at RAF High Ercall.

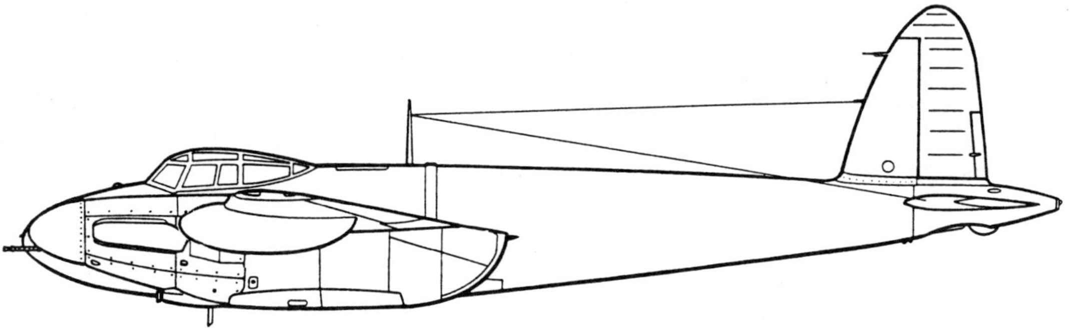

Mosquito FB.VI

DH-98 Mosquito NF.XVII (Jan 44 - Feb 45)

One hundred NF.II aircraft were converted to carry the American SCR720 radar, known as AI Mk X, and designated Mosquito NF.XVII. These aircraft were distinguished by the bulbous "universal" radome, which enabled the later marks of Mosquito to carry either AI Mk VIII or Mk X, and were powered by Merlin 23 or 25 engines. The first five were received on 456 Squadron on 29 January 1944, with the first operations being flown on 12 February 1944. At the end of the month, 456 was replaced at Fairwood Common by No 68 Sqn, and moved to replace 29 Sqn at the forward base of Ford, on the Sussex coast. Because of the high security of the new radar equipment, these aircraft carried serial numbers with a G suffix (eg HK246/G) to indicate the need for armed guarding at all times. In December 1944, 456 moved north to Church Fenton, Yorkshire, to replace a Polish night fighter unit, No 307 Sqn.

Serial	Date On	Code	Date Off	Fate
HK246	31.1.44	RX-U	15.2.45	to 51 OTU Cranfield
HK248	28.10.44		19.12.44	crashed landing Manston (Porter/Guest), caught fire and written-off
HK249	31.1.44	RX-B	1.2.45	damaged 20.8.44, to 51 OTU
HK250	27.10.44	RX-P	18.1.45	to 68 Sqn Coltishall (coded WM-Z)
HK253	29.1.44	RX-F	8.2.45	to 51 OTU
HK264	29.1.44	RX-T	8.3.45	damaged 12.12.44 when fired at by Wellington (Samson/Abbey), to 51 OTU
HK282	10.2.44	RX-L	18.1.45	to 51 OTU (ex VY-D of 85 Sqn)
HK286	29.1.44	RX-A	4.1.45	to 51 OTU and damaged 3.3.45
HK290	8.2.44	RX-J	18.1.45	to 51 OTU
HK295	31.5.44	RX-X	22.10.44	overshot runway at Ford (Gould/Fleming), written-off

Serial	Date On	Code	Date Off	Fate
HK297	3.2.44	RX-V	21.6.44	to 219 Sqn Bradwell Bay, then 51 OTU
HK302	4.3.44	RX-Z	8.2.45	to 51 OTU
HK303	6.2.44	RX-H	18.1.45	to 51 OTU then 54 OTU
HK304	4.11.44	RX-N	18.1.45	to 68 Sqn then 51 OTU
HK312	31.1.44	RX-G	12.7.44	crashed into sea off Littlehampton on anti-diver patrol (Radford/Atkinson killed)
HK313	31.1.44	RX-W	4.1.45	to 51 OTU, convtd to instr a/f 7.7.45 as 5286M
HK317	6.2.44	RX-Y	24.11.44	failed to return from anti-diver op (Mulhall/Jones killed)
HK321	8.2.44	RX-X	29.4.44	failed to return from anti-minelaying sortie off Isle of Wight (Pahlow/Silva killed)
HK323	10.2.44	RX-R	4.1.45	to 51 OTU
HK353	.4.44	RX-M	16.7.44	damaged 11.6.44 by He 177 (Sanderson/Nicholas), to 10 MU for repair
	28.10.44	RX-M	2.45	to 51 OTU, damaged 26.2.45
HK356	29.1.44	RX-D	4.1.45	to 51 OTU, damaged 5.2.45
HK357	27.10.44		7.11.44	failed to return from patrol to Belgium (Woodward/Madden killed)
HK359	3.2.44	RX-K	11.1.45	to 51 OTU
HK360	29.1.44		17.4.44	damaged on ground Fairwood Common 7.2.44, repaired on 456, to FIU, damaged 23.9.44

HK286 (RX-A) was the second highest scoring Mosquito night fighter in the RAF with seven victories. The last 456 Mosquito NF.XVII operation was flown on 4 February 1945, and most surviving aircraft were passed to No 51 OTU, at RAF Cranfield, Bedfordshire.

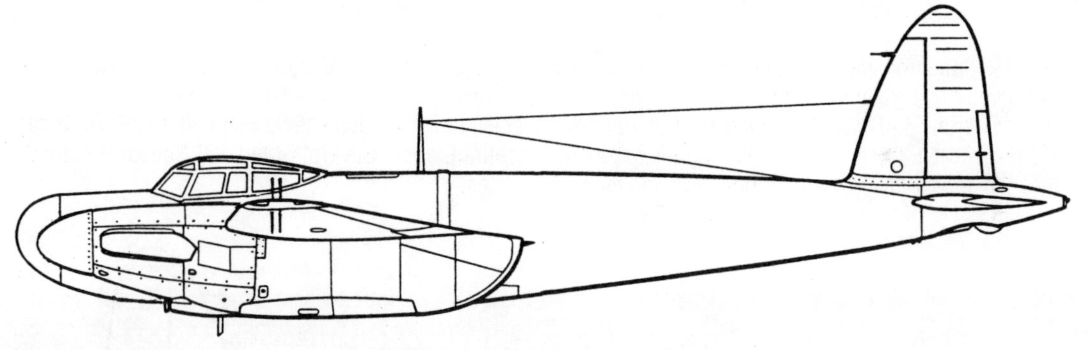

Mosquito NF.XVII

DH-98 Mosquito NF.30 (Dec 44 - Jun 45)

From December 1944, a high altitude Mosquito variant, the NF.30, powered by the 1680hp Merlin 72 or 1710hp Merlin 76 two-stage engines, was received on 456 Squadron. The cooling intake was fitted forward of the radiator to give these variants of the Merlin its distinctive lip beneath the spinner. The NF.30 was also fitted with the Mark X AI radar, and was the most numerous of the late war Mosquito night fighters, with 530 being built. It was capable of a maximum speed of 407mph, and a range of 1300 miles.

The first three NF.30s for 456 were delivered from No 96 Sqn at Odiham, Hampshire, on 7 December. Again, because of the sensitivity of the radar equipment that was carried, these aircraft were each serialled with the G suffix. The first op was flown by the new aircraft on 12 December 1944. After several months based at Church Fenton, in March 1945, 456 was replaced by No 68 Sqn, and moved once again. This time it was to the Essex coast base of Bradwell Bay, from where 456 operated with No 151 Sqn.

Serial	Date On	Code	Date Off	Fate
MM687	9.12.44	RX-Z	17.4.45	damaged landing Bradwell Bay 26.3.45 (Boorman/Morgan), to repair (Belgian AF 13.1.49 as MB-21)
NT241	16.12.44	RX-W,-X	5.7.45	to 44 MU, then 54 OTU and written-off 29.1.46
NT245	28.12.44		24.5.45	to 125 Sqn Church Fenton, then 25 Sqn (coded ZK-D)
NT264	7.12.44	RX-R	29.5.45	ditched on approach to Bradwell Bay (Howard killed)
NT269	7.12.44	RX-J	26.6.45	to 10 MU, then 57 MU
NT282	7.12.44	RX-T	5.7.45	to 54 OTU Charter Hall, then 502 Sqn
NT295	24.12.44	RX-P	5.7.45	to 54 OTU, damaged 27.8.45
NT296	11.1.45	RX-S	5.7.45	to 54 OTU, damaged 6.7.45
NT298	22.12.44		5.7.45	to 54 OTU, damaged 9.10.45
NT299	16.12.44	RX-D	14.2.45	crashed landing Woodvale, Lancs (Cowper) written-off
NT300	28.12.44		26.6.45	to 10 MU (to Belgian AF 23.1.48 as MB-4)
NT302	24.12.44	RX-K	26.6.45	damaged landing Bradwell Bay 22.4.45 (Johns/ Bates), repaired on 456, to 10 MU, then 274 MU
NT307	29.12.44	RX-Y	12.7.45	to 54 OTU, then 273 MU
NT311	28.12.44	RX-L	26.6.45	to 10 MU, then 616 Sqn
NT328	24.12.44	RX-H	26.6.45	to 10 MU
NT332	28.12.44	RX-B	26.6.45	to 10 MU (to Belgian AF 13.1.49 as MB-22)
NT335	14.1.45	RX-V	5.7.45	to 44 MU
NT355	25.1.45		28.6.45	to 10 MU
NT363	8.2.45	RX-R,-U	5.7.45	to 54 OTU, then convtd to instr a/f in 1946 as 6118M
NT387	20.2.45	RX-D	26.6.45	damaged landing Church Fenton 20.2.45 (delivery by ATA), repaired, 10 MU (to Belgian AF 16.9.48 as MB-16)
NT545	5.4.45	RX-Z	5.7.45	to 44 MU, then 54 OTU
NT602	1.6.45		26.6.45	to 10 MU (to French AF 15.12.48)

The last 456 Squadron wartime operation was flown by two Mosquito aircraft (NT295 and NT387) on 3 May 1945. The Squadron's flying commitment ended on 8 June, when all Mosquitos were stored pending issue to other units. No 456 Squadron RAAF was disbanded at Bradwell Bay a week later, on 15 June 1945. Postwar, the RAF provided many Mosquito NF.30s to the re-emerging air forces of liberated Europe. Several of No 456 Squadron's aircraft were to see service flying with the Belgian and French Air Forces.

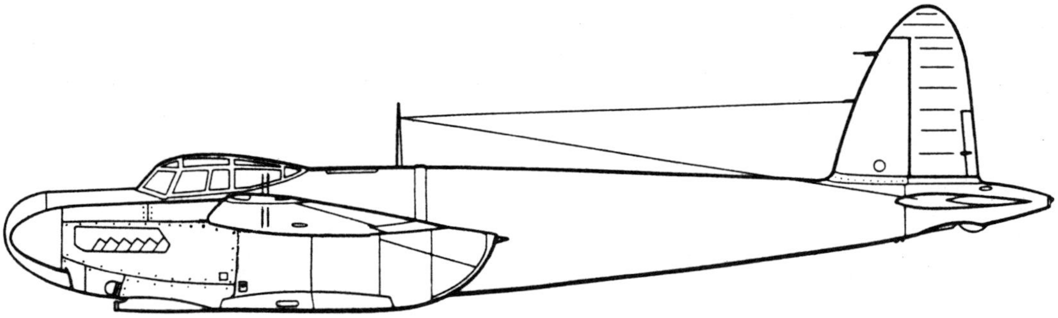

Mosquito NF.30

APPENDIX 8

456 CLAIMS AGAINST GERMAN AIRCRAFT

Date	Crew	Destroyed	Probable	Damaged	Remarks
10/11 Jan 42	Hamilton/N-Smith	1 Do 217E			F8+IN of 5/KG40[1] crashed Arbury, nr Nuneaton, Merseyside
18 May 42	Wills/Lowther	1 Ju 88D			convoy patrol east of Bardsey, 4U+FL of 3(F)/123[2] crashed into Cardigan Bay, Wales
29/30 Jul 42	Wolfe/Ashcroft	1 He 111H			F8+LW of 12/KG40 crashed onto Pwllheli beach, Lleyn Peninsula, Wales
7/8 May 43	Oxlade/Shanks	1 Ju 88A[3]			3E+KL of 3/KG6, dived into ground on intercept nr Hampshire/Dorset border
11 Jun 43	Newell/Keating			2 Ju 88C	Instep patrol, Bay of Biscay, V/KG40
19 Jun 43	Gatenby/Fraser			1 Ju 88C	Instep patrol, off Cape Ortegal, Bay of Biscay, V/KG40

1 The usual abbreviation for a Luftwaffe unit: 5/KG40 represents No 5 Staffel of Kampfgeschwader 40, or V/KG40, for V Gruppe of KG40.

2 A more unusual squadron designation: No 3 (Fernaufklärungs/ long-range) Staffel of reconnaissance Gruppe 123.

3 Originally the aircraft type was unknown and not credited, but was later identified as Ju 88A which crashed near Winfrith, Dorset, at 2325 hours.

Date	Crew	Destroyed	Probable	Damaged	Remarks
21 Sep 43	Panitz/Williams	1/2 Ju 88C			shared with Samson, Instep patrol, Bay of Biscay, V/KG40
	Samson/Abbey	1/2 Ju 88C			shared with Panitz
	Gatenby/Fraser		1 Ju 88C		Instep patrol, Bay of Biscay, V/KG40
	Newell/Keating			1 Ju 88C	Instep patrol, Bay of Biscay, V/KG40
3 Oct 43	Panitz/Williams	1 Ju 88C			Instep patrol, off Biscarrosse, Gironde
25 Feb 44	Samson/Abbey	1 Ju 88			Ranger patrol, Rennes airfield
1/2 Mar 44	Richardson/Landy			1 Do 217	over Sussex
21/22 Mar 44	Thornley/Phillips		1 Fw 190		off French coast
	Roediger/Dobson	1 Ju 88			in sea off Rye, East Sussex
24/25 Mar 44	Hampshire/Condon	1 Ju 88A			3E+AP of 6/KG6 crashed at Walberton, nr Ford, Sussex
27/28 Mar 44	Hampshire/Condon	1 Ju 88A			3E+FT of 9/KG6 crashed nr Beer, Devon
		1 Ju 88A			B3+BL of 3/KG54, crashed nr Isle Brewers, Somerset
	Howard/Ross	1 Ju 88			off Cherbourg coast
19/20 Apr 44	Brooks/Forbes	1 Me 410A			9K+JH of 1/KG51, crashed Nuthurst, south of Horsham, Sussex
23/24 Apr 44	Hampshire/Condon	1 Ju 88			off Swanage, Dorset, cannon failed on attacking 2nd Ju 88
25/26 Apr 44	Lewis/Wilmoth	1 Ju 188			nr Portsmouth
	Roediger/Dobson	1 Ju 188			in sea off Portsmouth
	Houston/Engberg	1 Ju 88			in sea off Portsmouth
28/29 Apr 44	Hampshire/Smith		1 Do 217[4]		probably from III/KG100 minelaying in English Channel 86 miles off Durrington
14/15 May 44	McEvoy/Austin	1 Ju 188A			U5+HH of 1/KG2, crashed Larkhill, nr Netheravon, Salisbury Plain, Wiltshire
15/16May 44	Arnold/Stickley	1 Ju 88A			B3+DT of 9/KG54, crashed at Medstead, Hampshire
22/23 May 44	Hampshire/Condon	1 Ju 88			off Southampton
	Sanderson/Nicholas	1 Ju 88			off Southampton
6/7 Jun 44	Pratt/Smith	1 He 177			probably KG40[5], off Normandy
	Stevens/Kellett	2 He 177			Cherbourg/Le Havre
	Hampshire/Condon	1 He 177			over sea east of Barfleur, Cherbourg Peninsula
7/8 Jun 44	Howard/Ross	2 He 177			missile launchers, off

4 The Do 217s claimed during 1944 were most probably from III/KG100, based at Toulouse-Francazal.

5 No 456 Sqn was the top scoring Mosquito unit against the He 177. All of the victims were most probably from KG40, based at Toulouse-Blagnac or Bordeaux-Mérignac.

Date	Crew	Destroyed	Probable	Damaged	Remarks
					Normandy
	Hodgen/McCormick	1 He 177			off Normandy
9/10 Jun 44	Cowper/Watson	1 He 177			off Cap Levy, originally claimed as probable, wreckage later found nr Cherbourg
		1 Do 217			from III/KG100, nr Cap de la Hague (either Do 217M 6N+JP or Do 217K 6N+KT)
10/11 Jun 44	Sanderson/Nicholas	1 He 177			English Channel, major wing damaged by return fire
	Howitt/Irving		1 He 177		seen going down over English Channel with engine on fire, 40 miles south of Brighton
12/13 Jun 44	Hampshire/Condon	1 Ju 88			believed to be night fighter variant, Caen Bay
13/14 Jun 44	Williams/Havord	1 He 177			mine-laying off Fecamp, nr Le Havre
	Thornley/Phillips			1 Hs 129[6]	later claim, after oil found covering front of Mosquito
14/15 Jun 44	Cowper/Watson	1 Ju 88			over sea nr Cherbourg Peninsula, 3 crew baled out
22/23 Jun 44	Stevens/Kellett		1 Ju 88G		night fighter, English Channel
4/5 Jul44	Sanderson/Nicholas	1 He 177			English Channel
	Cowper/Watson	1 He 177			English Channel, south of Selsey Bill
	Williams/Havord	1 Do 217			believed to be Do 217M 6457 of III/KG100, English Channel
	Radford/Atkinson	1 He 177			English Channel
6/7 Oct 44	Mulhall/Jones	1 Ju 188			20 miles north-west of Nijmegen, Holland
19/20 Nov 44	Arnold/Stickley	1 He 111H			V-1 launcher of KG53, in sea 75 miles east of Lowestoft
24/25 Nov 44	Stevens/Kellett	1 He 111H			V-1 launcher of KG53, 10 miles off Texel, Holland
4/5 Mar 45	Hodgen/McCormick		1 Ju 188		30 miles east of Spurn Head, Yorkshire

In addition, during the Squadron's intruder patrols of Munich's airfields in late April 1945, other claims were lodged against aircraft on the ground. On the night of 22/23 April, McWhinney/Costigan claimed one unidentified aircraft destroyed at Bad Aibling, and on the night of 25/26 April, Richardson/Landy claimed one unidentified aircraft damaged at Schleissheim.

6 No order of battle substantiates the existence of the Henschel Hs 129 "tank buster" in this theatre of operations. As the identification was from recognition charts during debrief, this type must remain in doubt.

APPENDIX 9

456 CLAIMS AGAINST V1 FLYING BOMBS

Date	Crew	Destroyed	Probable/ Damaged	Remarks
24/25 Jun 44	Houston/Engberg		1	off Beachy Head, 3500', 400mph, heaing 340, strikes observed, possible impact seen 3 minutes later
9/10 Jul 44	Roediger/Dobson	1		V1 2500' over English Channel, travelling 330 mph, heading 330
10/11 Jul 44	Houston/Engberg	1		V1 2000' over English Channel, travelling 300 mph, heading 340
13/14 Jul 44	Roediger/Dobson	1		V1 1500' over English Channel, travelling 320mph, heading 320
17/18 Jul 44	Sanderson/Nicholas		1	strikes seen on V1 but no crash seen
18/19 Jul 44	Roediger/Dobson	1		V1 300' over Kent, travelling 200 mph, heading 310-320
20/21 Jul 44	Pratt/Smith	1		V1 1200' over Kent, travelling 280 mph, heading 320
	Brooks/Forbes	1		V1 1200' over Kent, travelling 350 mph, heading 330
	Stevens/Kellett		1	V1 500', travelling 400mph+, strikes seen on V1 as it flew towards balloon barrage
27/28 Jul 44	Cowper/Watson	1		V1 800' over Sussex, travelling 280mph, heading 010, not seen to crash, later confirmed
	Stevens/Kellett	1		V1 500' over Kent/Sussex, travelling in excess of 440mph
	Stevens/Kellett	1		V1 700', travelling 430mph nr Tonbridge, Kent, dived into ground and exploded[1]

Date	Crew	Destroyed	Probable/ Damaged	Remarks
30/31 Jul 44	Lewis/Wilmoth	1		V1 1500' over Sussex, travelling 300mph, heading 340
4/5 Aug 44	Roediger/Dobson	1		V1 3500' over Sussex, travelling 350mph, heading 350
5/6 Aug 44	Semmens/Nitschke	1		V1 1800' over English Channel off Eastbourne, travelling 330 mph, heading 360
	Williams/Havord	1		V1 2000' over English Channel off Dungerness, travelling 350 mph, heading 330
6/7 Aug 44	Roediger/Dobson	1		V1 700' over land, travelling 380 mph, heading 350
7/8 Aug 44	Howard/Ross	2		V1 1000', travelling 360mph, heading 010 10 minutes later, V1 800', 370 mph, heading 360
8/9 Aug 44	Howitt/Irving	1		V1 600' over land, 340 mph, heading 350
9/10 Aug 44	Roediger/Dobson	1		V1 2500' over land, travelling at over 400mph, heading 350
11/12 Aug 44	Houston/Engberg	1		V1 2300' over English Channel, travelling 380mph, heading 340
	Howard/Ross	1		V1 200' over English Channel, travelling 360mph, heading 350
12/13 Aug 44	Pratt/Smith	1		V1 1000' over English Channel, travelling 350mph, heading 340
	Howitt/Irving	1		V1 1500' over English Channel, travelling 250mph, heading 350, damage to Mosquito's port wing
13/14 Aug 44	Williams/Havord	1		V1 1000' over English Channel, travelling 350mph, heading 350
14/15 Aug 44	Hodgen/McCormick	1		V1 1500' over English Channel off Rye,travelling 360mph, heading 350, climbed to 4000', then dived vertically
19/20 Aug 44	Roediger/Dobson	1		V1 1200' over Sussex/Kent, travelling 360mph, 350 degrees
23/24 Aug 44	Roediger/Dobson	1		V1 4000' over English Channel, travelling 380 kts, heading 310
		1		5 mins later, V1 1500' over English Channel, 380mph, heading 310
30/31 Aug 44	Howard/Condon	1		V1 2000' over English Channel off Cap Gris Nez, travelling 350 mph, heading 320, debris hit Mosquito
		1		6 mins later, V1 6000' off Cap Gris Nez, travelling 300mph, heading 320

Nine flying bombs were shot down by Flight Lieutenant Keith Roediger, a Mosquito pilot of No 456 Squadron, whose navigator was Flight Lieutenant Bob Dobson. They were the top scorers for the unit. No 456 Squadron opened its flying bomb account on the night of 24 June 1944, when Flight Lieutenant George Houston and Flying Officer Lars Engberg claimed a probable. The Squadron destroyed 29 flying bombs in the succeeding 50 days.

Two RAAF airmen, flying Tempests on No 3 Squadron RAF, Flight Sergeants D J MacKerras and H G Bailey, each destroyed eleven V1s, to make them the highest scoring Australians.[2]

---

1 Although some records state both Stevens' V1s were at 3000', heading 340 degrees, between 270 and 350 mph, he is adamant that both were below 1000', travelling in excess of 400 mph.

2 Herington, Vol IV, p.185.

APPENDIX 10

THE GEE SYSTEM[1]

The *Gee* system consisted in the reception by equipment in the aircraft of pulse transmissions from three ground stations situated on a base line of approximately two hundred miles in length. One of these transmitters was known as the "A" or "Master" station and the other two were called the "B" and "C" or "Slave" stations. Each "Slave" transmission was locked to a "Master" transmission and the differences in the time taken by the "A" and "B" and the "A" and "C" signals to reach the aircraft were measured and displayed on a cathode-ray tube on the navigator's table in the aircraft. From them, the aircraft could be located on two position lines known as *Gee* coordinates and the ground position of the aircraft coincided with the point at which these coordinates, which were printed as a grid on special *Gee* charts, intersected. The coordinates could be obtained by a competent navigator in less than a minute.

The accuracy of a *Gee* fix varied from less than half a mile to about five miles and depended upon the skill of the navigator, upon whether the aircraft was in an area where the *Gee* grid intersected acutely or obtusely and upon the range of the aircraft from the transmitters. The range of *Gee* itself varied with conditions from three to four hundred miles, but in general, the rule was that the greater was the range the less was the accuracy.

Gee was thus an admirable aid to dead-reckoning navigation, for it provided a reasonably accurate and regular means of fixing the ground position. Over short distances it could also be used as a homing device. In this event, the *Gee* coordinates of the place it was desired to reach were set up on the cathode-ray tube and the aircraft was then steered in such a way as to bring the pulses into line. Originally it was hoped that this would offer a means of blind bombing but, in practice, as was shown by the 1942 attacks on Essen, *Gee* was not, at that range, sufficiently accurate for the purpose. Nevertheless, over England on the return flight, *Gee* was generally used for homing to base, which meant that aircraft could come home on prearranged coordinates without fear of getting lost and with a much reduced risk, while losing height, of running into hills or other aircraft.

LIMITATIONS AND DISADVANTAGES OF *GEE*

The principal limitation of *Gee* was its lack of range, to which reference has already been made. Over Germany, it was seldom of any use to the east of the Ruhr. In addition it was, as has also been mentioned, not accurate enough for effective blind bombing. The principal disadvantage of the system was its extreme susceptibility to jamming, which had been foreseen from the beginning. Almost exactly as had been expected, the Germans began effective jamming some five months after the general introduction of *Gee*, and after that fixes could rarely be obtained anywhere east of the Dutch coast, and Gee became useful only for navigation over the North Sea and

1 *The Strategic Air Offensive Against Germany 1939-1945,* Vol IV, Sir Charles Webster & N Frankland, HMSO, London, 1961, pp.4-6.

parts of France, as well as for homing to base. Another disadvantage of *Gee* was that, being more accurate over England than over Germany, it conferred a greater target-finding advantage on the *Luftwaffe* than on Bomber Command. This eventually led to the necessity for coding the *Gee* pulses, which increased the time taken to get a fix and raised posibilities of confusion which sometimes occurred.

DEVELOPMENT AND INTRODUCTION OF *GEE*

Scientific proposals for a system like that eventually used for *Gee* were put forward in 1938, but it was not until 1940 that the actual development of the apparatus was begun. Two officers from Bomber Command were given an airborne demonstration of the equipment by the Telecommunications Research Establishment in October 1940 and official recognition of the potential value of *Gee* followed. There were, however, differences of opinion as to whether the equipment should be put into the earliest possible operational use in a small number of bombers which could be used in a kind of pathfinder role as fire-raisers or whether it should be delayed until a really substantial number of bombers could be equipped.

While this debate was going on, the service trials of *Gee* were begun in May 1941 when the device was still primarily regarded as a means of blind bombing, but mass production was delayed by severe technical difficulties with the production of one of the valves. Nevertheless, a small number of Wellingtons belonging to 115 Squadron had been equipped with *Gee* by July 1941 and the system was operated over Germany for the first time in two of these Wellingtons on the night of 11th August 1941 when very encouraging results were obtained over the Ruhr. On the night of 12th August two *Gee* Wellingtons flew over Hanover, but one of them failed to return, and after a third *Gee* operation, also involving the use of two Wellingtons, it was decided not to make any further use of the equipment until three hundred bombers could be equipped with it. This target had, however, by no means been reached, when on the night of 8th March 1942 *Gee* was put into general operational service with Bomber Command. Though there is no evidence to suggest that the Germans learnt anything from the *Gee* Wellington which was brought down on the night of 12th August 1941, they did start effective *Gee* jamming on the night of 4th August 1942.

Despite the jamming, *Gee* justly continued to be regarded as essential equipment for Bomber Command. By August 1942, eighty per cent of the operational force was equipped with it and by January 1943 the whole force had been fitted. The *Gee* coverage, which had initially been focused on the Ruhr by a group of transmitters known as the Eastern Chain, was also increased by the opening of further chains among which were the Northern and Southern Chains, the latter of which covered northern France. An improved version of the equipment, known as *Gee* Mark II, was operational introduced in February 1943, and by the end of March 1943, sixty per cent of the force had received the new equipment. Its anti-jamming devices afforded some slight and temporary releif from the German counter-measures, but apart from being easier to use in the air, the Mark II equipment had no other significant advantages by comparison with the Mark I. No effective and lasting means of cutting out the enemy jamming of *Gee* was ever found, but throughout the war the equipment was always valued by those who used it even if only because it offered them an easy and safe passage home after they had completed their operation.

APPENDIX 11

ROGUES' GALLERY

A collection of charactertures of 456 Squadron personalities, drawn by Sergeant A Barnett, of RAAF Overseas Headquarters, London, in 1942.

APPENDIX 11

ROGUES GALLERY

A collection of charactertures of 456 Squadron personalities, drawn by Sergeant A Barnett, of RAAF Overseas Headquarters, London, in 1942.

PERSONALITIES
BY
A. BARNETT
403330. F/S. GATENBY. G.F
(PILOT)
W/CMDR.
G.C. OLIVE
D.F.C.
(R.A.F)
401343
SGT.
WOOD. G.A.C
(OBSERVER
403654
SGT.
KEATING A
(OBSVR)
405183
SGT.
McCORMACK. L.M
(OBSERVER
403373
SGT
RATCLIFF V.P
PILOT
404220.
SGT.
STOVELL. C.

406279
SGT
GILL.L
(OBSV)
403115
SGT
SKIPPEN L.T
(OBSERVER)
SGT
FRASER J.M
(OBSV)
SGT
HUGHES.T
(A.G)
F/SGT
SCOTT.R.M
(PILOT)
403821
P/O
SMITH.H
(PILOT)
400739
F/SGT
STEVENS.F.S
(PILOT)
404394
P/O
CONDON.T
(OBSV)

402816
F/SGT
SPRING. D A H
404543
SGT
(PILOT)
400725
P/O
400736
W/O
(PILOT)
1018
PIKE
(ADJT)

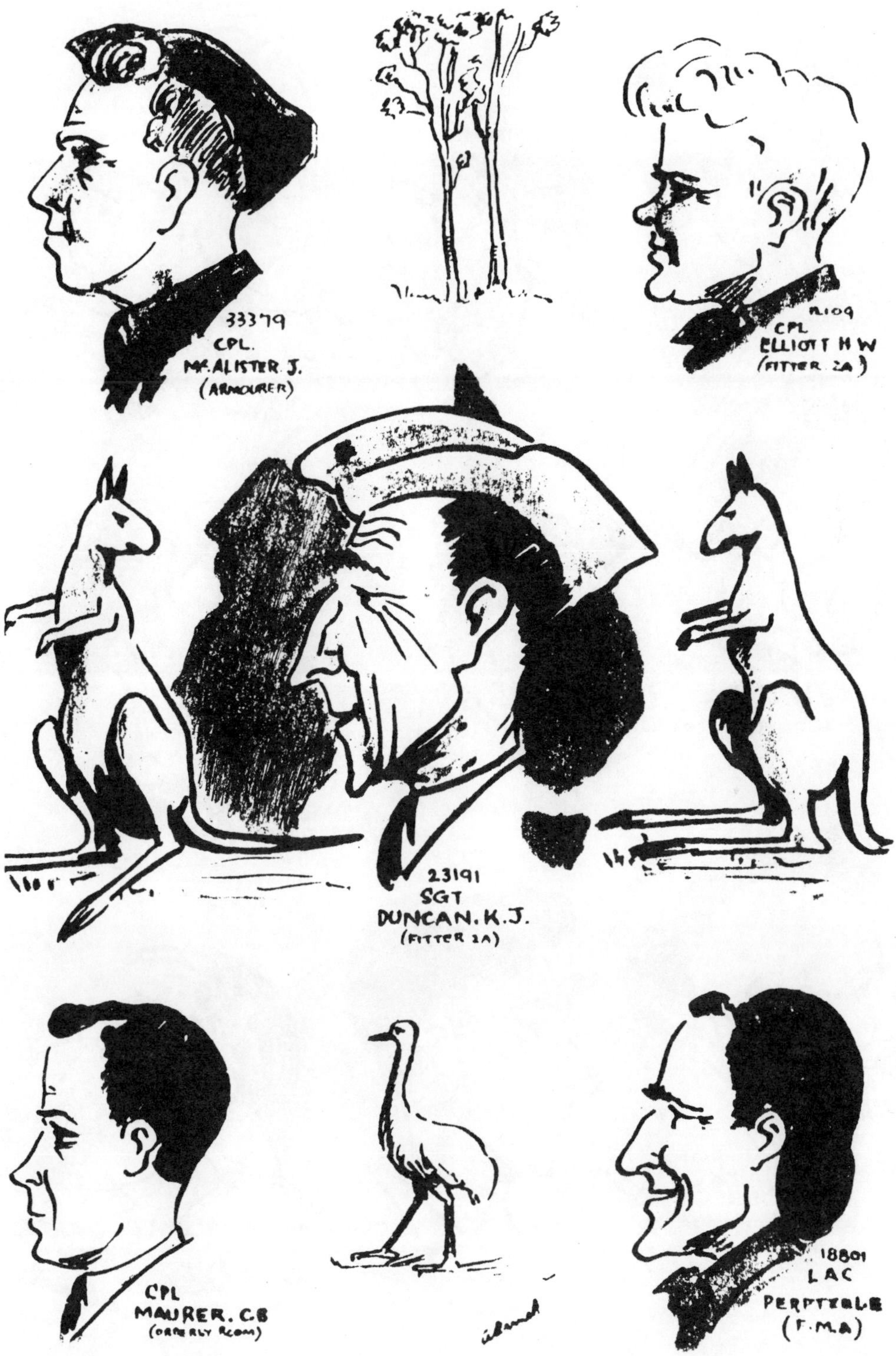
33379
CPL.
McALISTER. J.
(ARMOURER)
R109
CPL
ELLIOTT H W
(FITTER 2A)
23191
SGT
DUNCAN. K. J.
(FITTER 2A)
CPL
MAURER. C.B
(ORDERLY ROOM)
18801
LAC
PERPTEBLE
(F.M.A)

33433
L A C
BENNING J.R
20598
L.A.C.
CHAPMAN O.K
(ARMOURER)
8998
L.A.C
ALLISON.C.A
26125
F/S.
MACKIE. W.
(DISCIP)
16512
CPL
THOMPSON.W
5633
SGT
de'SALIS F
18371
L.A.C
WILSON. D
L A C
STAMP. L.
L.A.C
A Barnett.

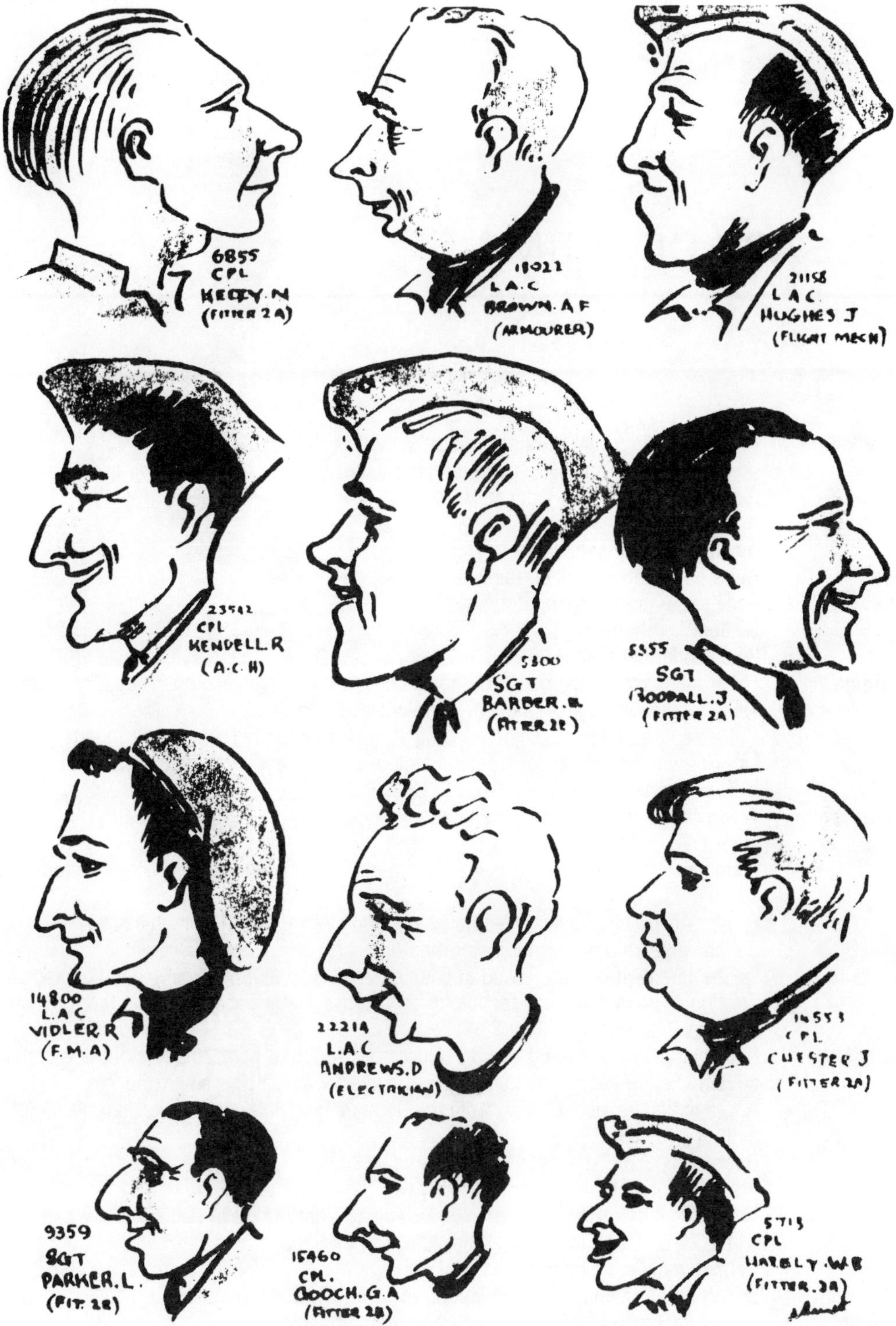
6855
CPL
KELLY.N
(FITTER 2A)
18022
L.A.C
BROWN.A.F
(ARMOURER)
21158
L.A.C
HUGHES.J
(FLIGHT MECH)
23542
CPL
KENDELL.R
(A.C.H)
5300
SGT
BARBER.R
(FITTER 2E)
5355
SGT
GOODALL.J
(FITTER 2A)
14800
L.A.C
VIDLER.R
(F.M.A)
22214
L.A.C
ANDREWS.D
(ELECTRICIAN)
14553
CPL
CHESTER.J
(FITTER 2A)
9359
SGT
PARKER.L
(FIT. 2E)
15460
CPL
GOOCH.G.A
(FITTER 2E)
5713
CPL
HATELY.W.B
(FITTER 2A)

GLOSSARY OF TERMS

AA, or Ack-Ack	Anti-aircraft fire from the ground, also known as flak
ADGB	Air Defence of Great Britain, the temporary renaming for the RAF's Fighter Command after the forming of the 2nd Tactical Air Force in 1944
AI	Airborne Interception radar
AOC	Air Officer Commanding
Bandit	Airborne target, almost certainly an enemy
Big Ben	Code name for anti V2 operations
Bogey	Unidentified target
C-in-C	Commander-in-Chief
Chain Home	The British radar network established along the eastern and southen coastlines from 1939, to provide early warning of German air attacks
Circus	Fighter-escorted light daylight bombing attacks against short-range or fringe targets to lure German fighters to battle, and maintain a strong enemy fighter force on the particular front concerned
Detachment	A part of a unit, detached from its parent unit to another base for operations in another area
Diver	Code name for the V1 flying-bomb - operations to shoot down V1s became anti-diver patrols
EATS	Empire Air Training Scheme, established during World War II to train aircrew in Australia, Canada and Rhodesia primarily for the RAF
FIU	Fighter Interception Unit, based at Ford, charged with testing experimental equipment and developing procedures and tactics for service use - later became Fighter Interception Development Squadron (FIDS)
Fighter Night	Code name for a system, started in November 1940, for fighter patrols above a certain height, at night, when any twin-engined aircraft was considered hostile
Flak	Anti-aircraft fire, abbreviation from the German *Fliegerabwehrkanone* (aeroplane defenc gun)
Flash	Operate the AI
Flight	A sub-division of a squadron, normally two flights to a squadron
Flower	Code name for intruder operations over enemy night fighter airfields in order to keep them on the ground, or intercept any that took off or landed
Freya	German early warning radar
GCI	British Ground Controlled Interception radar for controlling fighters
Himmelbett	*Luftwaffe* procedure for ground control of night fighters
IFF	British encrypted Identification Friend or Foe equipment, which retransmitted the signal of friendly AI or GCI radars

Instep	Patrols over the western approaches and Bay of Biscay, to intercept German fighters inter fering with Coastal Command operations
Jim Crow	Operational patrols of the home coastline, initially to supplement the information provided by the Chain Home radars, to intercept any hostile aircraft crossing the coast
Judy	Code word signifying satisfactory AI contact to contine the interception alone
Kammhumer Line	German line of night defence, named by the British after the *Luftwaffe* "General of Night Fighters", to protect the Ruhr industries, by rings of AA batteries and lines of searchlights, with a night fighter zone stretching from Kiel to Liege
Knickebein	German for "crooked leg" - the *Luftwaffe* radio beam system that enabled "blind" bomb ing over England
Lichtenstein	The first German AI radar - the later model of which was designated SN2
Mahmoud	Code name for British bomber support operations by Mosquitos fitted with rear-facing AI radar, along the bomber streams or at German night fighter assembly points - the intent being when a target was located, a 360 degree turn was executed to get onto the enemy's tail
Monica	British rearward facing AI which acted as a tail warning radar
Noball	Code name for launching, storage and manufacturing sites for German V-weapons
OTU	Operational Training Unit
Ramrod	Code name for operations similar to a Circus but with a primary aim of destruction of a target
Ranger	Code name for operations, usually of squadron, wing or group strength, aimed as free lance intrusions over enemy territory with the aim of wearing down the enemy fighter force
Rhubarb	Small-scale fighter or fighter-bomber attacks on ground targets of opportunity
Roadstead	Code name for operations ships by fighters or bombers escorted by fighters
Rodeo	Code name for fighter sweeps over enemy territory
Rover	Code name for armed reconnaissance against chance targets behind enemy lines
Schräge Musik	literally "slanting music", the twin 20mm obliquely upward-firing machine guns installed in *Luftwaffe* night fighters
Scramble	Calling aircraft on readiness into the air to intercept hostile or unidentified intruders
Serrate	Code name for equipment to enable homing onto the enemy's AI transmissions - combined with AI on Mosquitos to attack German night fighters
Section	Sub-division of a Flight, comprising two to four aircraft and defined by a colour, eg Blue Section
Sortie	An aircraft's individual operational flight - if the Squadron launched twelve aircraft on a sweep, twelve sorties would be flown on that one operation
Standing Patrol	A patrol maintained at set times, whether the presence of enemy forces is anticipated or not
Sweep	An offensive formation of fighters or fighter bombers designed to draw the enemy
USAAF	United States Army Air Force
Wilde Sau	*Luftwaffe* "wild boar" tactics for night fighters, operating above the flak and searchlight defences, attempting intercepts from a ground contol "running commentary" of all con tacts
Window	Code name for strips of tinfoil, dropped from RAF aircraft to stimulate returns on enemy radars to represent aircraft echoes and confuse radar operators - now known as "chaff"
Würzburg	German AA radar, adapted as a GCI for the contol of fighters
VHF	Very high frequency radio band, between 30 MHz and 300 MHz, used for communications and navigation aids
V Weapon	Hitler's reprisal weapons, the V1 rocket bomb V2 guided missile. From the German *Vergeltungswaffe*
X-Gerät	X-apparatus equipment carried by *Luftwaffe* aircraft, for use in conjunction with the *Knickebein* navigation system, utilising two transmitted ground radar beams - superceded by single-beam *Y-Gerät* equipment

BIBLIOGRAPHY

PRIMARY SOURCES

Air Ministry Form 78, Aircraft Status Cards, RAF Air Historical Branch, Ministry of Defence, London.
Air Ministry Form 1180, Accident Cards, RAF Air Historical Branch, Ministry of Defence, London.
Casualty Registers, RAAF Historical Section, Canberra.
Combat Reports, RAAF Historical Section, Canberra
Honours and Awards Register, RAAF Historical Section, Canberra
Operational Record Book, Form 540, No 456 Squadron, RAAF Historical Section, Canberra.
Record of Operations, Form 541, No 456 Squadron, RAAF Historical Section, Canberra.
Personal Interviews and Letters
RAAF Personnel Records, Personnel Records Section, Queanbeyan

SECONDARY SOURCES

Air Ministry, *The Rise and Fall of the German Air Force*, Air Ministry Pamphlet No 248, HMSO, London, 1948
Austin, M N, *Night Fighters and the History of No 456 Squadron RAAF*, unpublished manuscript, Air Ministry, London, 1945, RAAF Historical Section, Canberra
Barnes, C H, *Bristol Aircraft Since 1910*, Putnam, London, 1994.
Bartz, K, *Swastika in the Air*, Kimber, London, 1956.
Bekker, C, *The Luftwaffe War Diaries*, MacDonald, London, 1966.
Bennett, J W, *Defeat To Victory*, RAAF Museum, Point Cook, Vic, 1994.
Bowyer, C, *Fighter Command 1936-1968*, Sphere, London, 1981.
Charlton, ACdre L E O, *Britain At War - The Royal Air Force*, April 1942 - June 1943, Hutchinson, London, 1945.
Churchill, W S, *The Second World War, Vol VI*, Cassell, London, 1954
Clostermann, P, *The Big Show*, Vanguard, London, 1955.
Collier, B, *The Defence of the United Kingdom*, HMSO, London, 1957.
Franks, N, *RAF Fighter Command*, Haynes, Yeovil, Somerset, 1992.
Gunston, B, *Night Fighters*, Patrick Stephens Ltd, Cambridge, 1976.
Heeks, K, *Bombing 1939-45*, Hale, London, 1990.
Howard-Williams, J, *Night Intruder*, Airlife, Shrewsbury, 1992.
Herrington, J, *Air War Against Germany and Italy 1939-1943*, RAAF Official History Vol III, AWM, Canberra, 1962.
Herrington, J, *Air Power Over Europe 1944-45*, RAAF Official History Vol IV, AWM, Canberra, 1963.
Hinsley, F H, *et al*, *British Intelligence in the Second World War*, Vol III, Part II, HMSO, London, 1988
Jackson, A J, *De Havilland Aircraft Since 1909*, Putnam, London, 1994.
Johnson, F, *RAAF Over Europe*, Eyre & Spottiswoode, London, 1946.
Johnson, AVM J E, *The Story of Air Fighting*, Hutchinson, London, 1985.
Kelly, J W, *The Wooden Wolf*, Futura, London, 1978.
Martin, D, *50 Years in Radar*, Thorn EMI, Haynes, 1986.
Middlebrook, M, & Everitt, C, *The Bomber Command War Diaries*, Penguin, London, 1990.

Ramsey, W G, editor, *The Blitz Then and Now*, Vol 3, Battle of Britain Prints Intl, London, 1990.
Rawnsley, C F, & Wright, R, *Night Fighter*, Collins, London, 1957.
Rawlings, J D R, *Fighter Squadrons of the RAF and their Aircraft*, MacDonald Janes, London, 1978.
Richards, D, *Royal Air Force 1939-1945*, Vol I, HMSO, London, 1953.
Richards, D, & Saunders, H St G, *Royal Air Force 1939-1945*, Vol II, HMSO, London, 1954.
Robertson, B, Airborne Interception, in *Aviation News*, 31 Dec 1993 - 27 Jan 1994.
Saunders, H StG, *Royal Air Force 1939-1945*, Vol III, HMSO, London, 1954.
Scutts, J, *Mosquito in Action Part 2*, Squadron/Signal, Carrollton, Texas, 1993.
Sharp, C M, & Bowyer, M J F, *Mosquito*, Faber, London, 1971.
Sims, E H, *Fighter Tactics and Strategy 1914-1970*, Cassell, London, 1972.
Smith, D J, *De Havilland Mosquito Crash Log*, Midland Counties, Earl Shilton, Leics, 1980.
Stevens, F S, *An Aussie in a Mossie*, unpublished manuscript, 1991.
Streetly, M, *Confound and Destroy*, MacDonald and Janes, London, 1978.
Thompson, G R, *United States Army in World War II, The Technical Services - The Signal Corps*, Dept of the Army, Washington, 1957.
Vincent, D, *Mosquito Monograph*, Self-published, Adelaide, 1982.
Webster, Sir Charles, & Frankland, N, *The Strategic Air Offensive Against Germany 1939-1945*, Vol IV, HMSO, London, 1961

INDEX